UNFINISHED BUSINESS
Telecommunications
after the Uruguay Round

GARY CLYDE HUFBAUER
ERIKA WADA
Editors

UNFINISHED BUSINESS
Telecommunications
after the Uruguay Round

INSTITUTE FOR INTERNATIONAL ECONOMICS
Washington, DC
December 1997

Gary Clyde Hufbauer is currently on leave as the Maurice R. Greenberg Chair and Director of Studies at the Council of Foreign Relations. He was formerly Reginald Jones Senior Fellow at the Institute for International Economics, Marcus Wallenberg Professor of International Finance Diplomacy at Georgetown University (1985-92), Deputy Director of the International Law Institute at Georgetown University (1979-81); Deputy Assistant Secretary for International Trade and Investment Policy of the US Treasury (1977-79); and Director of the International Tax Staff at the Treasury (1974-76). He has written extensively on international trade, investment, and tax issues, including *Flying High: Liberalizing Civil Aviation in the Asia Pacific* (1996), *Fundamental Tax Reform and Border Tax Adjustments* (1996) *Western Hemisphere Economic Integration* (1994), *Measuring the Costs of Protection in the United States* (1994), *NAFTA: An Assessment* (rev. 1993), *US Taxation of International Income* (1992), *North American Free Trade* (1992), and *Economic Sanctions Reconsidered* (second edition 1990).

Erika Wada is a Ph.D. candidate at Michigan State University and research assistant at the Institute for International Economics.

INSTITUTE FOR INTERNATIONAL ECONOMICS
11 Dupont Circle, NW
Washington, DC 20036-1207
(202) 328-9000   FAX: (202) 328-5432
http://www.iie.com

C. Fred Bergsten, *Director*
Christine F. Lowry, *Director of Publications*

Typesetting by Sandra F. Watts
Printing by Automated Graphic Systems

Printed in the United States of America
00 99 98   5 4 3 2 1

Library of Congress Cataloging-in-Publication Data

Unfinished business : telecommunications after the Uruguay Round / edited by Gary Clyde Hufbauer and Erika Wada.
    p. cm.
Includes bibliographical references.
    1. Telecommunication policy—Case studies. 2. Competition, International—Case studies. I. Hufbauer, Gary Clyde. II. Wada, Erika.
HE7645.U56 1997
384'.041—dc21   97-48348
                                           CIP
ISBN 0-88132-257-1

The views expressed in this publication are those of the authors. This publication is part of the overall program of the Institute, as endorsed by its Board of Directors, but does not necessarily reflect the views of individual members of the Board or the Advisory Committee.

Contents

IV Case Study

Appendix

Preface

Telecommunications liberalization is underway around the world. Two major agreements in 1997, the Telecom Pact in the World Trade Organization and the Information Technology Agreement (ITA), have set the pace at the international level.

However, questions about the wisdom of these actions are being raised in many countries. Efforts to introduce telecommunications competition in the United States through the US Telecom Act of 1996, for example, face many obstacles. Some analysts wonder if telecommunications, especially services based on physical networks, remain a natural monopoly or at least a natural oligopoly. Others fret about the pace of competitive pressure from large international telecom operators.

The authors in this volume argue that, to the contrary, vigorous competition is both possible and desirable and that effective regulation and enforcement can ensure competitive markets open to all firms. The volume brings together top telecommunications experts from a number of countries to identify industry trends, clarify issues, and make recommendations for future implementation of the WTO Telecom Pact, which is scheduled to begin taking effect in January 1998, and negotiations for an ITA II. They examine the recent trade agreements, assess their prospects, and call attention to new obstacles.

The Institute for International Economics is a private nonprofit institution for the study and discussion of international economic policy. Its purpose is to analyze important issues in that area and to develop and communicate practical new approaches for dealing with them. The Institute is completely nonpartisan.

The Institute is funded largely by philanthropic foundations. Major institutional grants are now being received from The German Marshall Fund of the United States, which created the Institute with a generous commitment of funds in 1981, and from The Ford Foundation, The Andrew W. Mellon Foundation, and The Starr Foundation. A number of other foundations and private corporations also contribute to the highly diversified financial resources of the Institute. The Pew Charitable Trusts provided support for this study. About 12 percent of the Institute's resources in our latest fiscal year were provided by contributors outside the United States, including about 6 percent from Japan.

The Board of Directors bears overall responsibility for the Institute and gives general guidance and approval to its research program— including identification of topics that are likely to become important to international economic policymakers over the medium run (generally, one to three years), and which thus should be addressed by the Institute. The Director, working closely with the staff and outside Advisory Committee, is responsible for the development of particular projects and makes the final decision to publish an individual study.

The Institute hopes that its studies and other activities will contribute to building a stronger foundation for international economic policy around the world. We invite readers of these publications to let us know how they think we can best accomplish this objective.

C. FRED BERGSTEN
Director
December 1997

Acknowledgments

We would especially like to thank Tomohiko Asano, Yuichi Takahashi, Guatam Jaggi, and Gena Morgan for their careful research assistance. We would also like to thank the publications staff, Brigitte Coulton, Kara Davis, Helen Kim, David Krzywda, and Christine Lowry, who carefully edited this volume and ensured rapid publication. Throughout the process, C. Fred Bergsten has provided constant guidance and support.

INTRODUCTION

1

Policy Issue Beyond the WTO Agreement

GARY CLYDE HUFBAUER AND ERIKA WADA

Twenty years ago, bringing competition to the telecommunications industry seemed impossible. The industry had been protected from competition by a massive wall built with official regulations, complex bilateral agreements, and concentrated market power. In 1994, the General Agreement on Trade in Services (GATS) was ratified with the purpose of eliminating that wall and other obstructions to trade in services. When GATS was established, agreement on liberalization of basic telecommunications services seemed a distant goal. However, member nations have treated the telecommunications sector as the prime target for liberalization, because they believe that telecommunications competition will facilitate the liberalization of other service sectors, including banking, consulting, and accounting. Far-reaching negotiations continued during and after the Uruguay Round, and on 15 February 1997, 69 countries that account for about 95 percent of the world's telecommunications traffic signed the World Trade Organization (WTO) Telecom Pact. This pact is the first multilateral agreement geared toward global liberalization of the telecommunications industry. However, the agreement—to be implemented beginning January 1998—is only the beginning of the battle to bring competition to the telecommunications industry. Many obstacles must still be overcome before competition becomes a practical

Gary Clyde Hufbauer was the Reginald Jones Senior Fellow at the Institute for International Economics. He is now the Maurice R. Greenberg Chair and director of studies at the Council on Foreign Relations. Erika Wada is a research assistant at the Institute for International Economics.

reality. This volume examines the telecommunications industry to see where it stands today on two questions: What did the recent trade agreements accomplish? And what does the larger battle look like?

The first part of this volume provides background on the WTO Telecom Pact agreement and an overview of its potential impact. The second part focuses on different aspects of the telecommunications industry: international trade in telecommunications services, the business practices of telecommunications firms, and the role of the telecommunications industry as the vital center of service industries. The final chapter and the appendices review the development of the telecommunications industry in the Asia Pacific Economic Cooperation (APEC) group.

The telecommunications industry has a unique history, because in the twentieth century it was perceived as a natural monopoly. In the era of simple copper cabling and electromechanical switchboards, competition seemed inefficient and impossible. As a result, monopoly telecommunication firms tied to national governments were the rule, in an effort to avoid price gauging and to ensure universal service. Competition was a rare exception. Indeed, most telecommunication firms were entirely government owned. Only recently have governments begun to privatize their national telecommunications firms, thereby separating operators (i.e., telecommunication firms) and regulators (i.e., specialized government agencies).

Technological progress and a global change in attitudes have led to much greater competition in the telecommunications sector. In a few industrial countries, telecommunications networks that previously carried mostly voice communications now offer a variety of services, which involve data, video, and computer-generated communications. Indeed, in rich countries of the Organization for Economic Cooperation and Development (OECD), data will soon surpass voice as the most significant telephone traffic. Many new services being embraced worldwide, such as wireless, the internet, and pagers, are supplied competitively rather than by monopolies.

In developing countries, wireless communications systems are beginning to compete with traditional wired systems in supplying basic telephone service. Moreover, advanced computer and communications technology have introduced indirect competition worldwide. For example, the emerging callback system threatens the monopoly profits of dominant carriers in many countries; email is replacing facsimile; and intranet is reducing the international phone bills of multinational corporations. Although consumers in many countries cannot yet choose their local or long-distance carriers, they can choose the mode of communication: phone, fax, or email. Pressure from this indirect competition has already forced traditional telephone companies to lower the price of some basic telephone services.

However, many of these alternative modes of communication still need telephone lines and networks to operate, and local exchange operators

(LECs) control this infrastructure. Consequently, although indirect competition has lowered the prices for some telecommunications services, traditional telephone companies still dominate the market. This market structure is not likely to change over the next decade. That is why many observers believe that government rules are necessary to bring competition to the telecommunications industry.

Jonathan D. Aronson's first contribution to this volume, Telecom Agreement Tops Expectations (chapter 2), examines the process of reaching the WTO Telecom Pact as well as the substance of the pact. Aronson argues that the telecommunications industry has entered a period of transition from national to global systems. Many countries with advanced telecommunications systems, including the United States, the United Kingdom, and Singapore, are trying to change the industry's market structure to expand their scope from national boundaries to global dimensions. Meanwhile, countries with less-advanced telecommunications systems are eager to acquire new technology to create a competitive infrastructure before they are overcome by foreign telecommunications firms. The WTO Telecom Pact reflected the different objectives of signatory nations. Because of US insistence on competitive safeguards that are actively promoted by national regulatory bodies, the WTO Telecom Pact was the first multilateral agreement to adopt competitive safeguards in industrial and developing countries. Aronson concludes that the WTO Telecom Pact is only the initial step toward a global and competitive industry.

What should be the role of the WTO in the long-term process of liberalization? What role should the other actors involved—national governments, telecommunications firms, and telecommunications customers —play in the new environment? Chapter 3 of this volume, Assessing the WTO Agreement on Basic Telecommunications, by William J. Drake and Eli M. Noam, attempts to answer these questions by laying out the potential benefits and costs of the WTO agreement.

In the first part of the chapter, Drake states his belief that the WTO agreement will significantly accelerate telecommunications liberalization, if effectively implemented. Opening existing markets is a short-term and necessary goal; over the longer term, in Drake's view, the WTO agreement will shape institutional governance. Drake also lists several reasons why the WTO agreement can lead to changes in institutional governance over the long term. The WTO agreement commits the telecommunications industry to the trade disciplines of the WTO. This implies that General Obligations and Disciplines (GODs) will apply to the telecommunications industry, just as they do to other trade-related industries. In addition, 48 countries signed an accompanying Reference Paper, which covers six main principles: Competitive Safeguards, Interconnection, Universal Service, Public Availability of Licensing Criteria, and Allocation and Use of Scarce Resources. Signatory governments promise to reform national regulatory rules to conform with the paper's key principles. The WTO's newly established Dispute Settlement Body (DSB) will monitor implementation of the agreement and, in Drake's view, give it teeth.

In the second part of chapter 3, Noam, in contrast, argues that although the WTO agreement is useful, its significance is greatly exaggerated. He says that most policy changes to foster competition and foreign investment would have taken place anyway. The WTO agreement, because it is complex and highly institutionalized, may even slow liberalization. For example, bringing a case to the WTO requires sophisticated legal and economic skills. And because anticompetitive practices are neither readily detected nor easily proven, vigilant national regulators are essential to realizing the potential gains from the WTO agreement. Such national regulators will emerge, in Noam's view, from national interests, not from WTO procedures.

With the conclusion of the WTO Telecom Pact, a looming challenge is to foster standardized interface technology for connecting telecommunications systems. In chapter 4, Telecommunications Liberalization: The Goods and Services Connection, John Sullivan Wilson suggests that the signing of the WTO Telecom Pact and the Information Technology Agreement (ITA) were initial steps toward the ultimate goal of a market-driven Global Information Infrastructure (GII). Wilson believes that new trade agreements will be required to reach that goal.

Wilson outlines how, under the ITA agreement, signatory countries promised to reduce tariffs on information technology (IT) goods to zero by 2000 or shortly thereafter. Wilson points out that, as tariffs decrease globally, the regulatory barriers to IT trade are becoming more conspicuous. Therefore, he argues that the next step is to reduce nontariff barriers. For example, national standards for electromagnetic compatibility, electrical safety, and telecommunications compatibility differ modestly from country to country. However, even small regulatory differences significantly raise production costs by raising compliance and certification costs.

The United States and European Union Mutual Recognition Agreement (MRA) talks in May 1997 focused on product testing and certification and represented a first effort to reconcile differing standards. The MRAs ensure safety, reliability, and interconnection while simplifying national certification processes. However, they are not easily negotiated, especially when several sectors are the subject of parallel talks. Wilson believes that other models for reaching agreement will be required and suggests that regional group such as APEC devise new ways to remove technical barriers.

While the issues discussed so far are relevant to other liberalizing sectors, such as financial services, the liberalization of telecommunications services brings about a specific set of complex issues. The next part of this book, Different Aspects of the Telecommunications Industry, deals with several of these sector-specific issues. Liberalization of the telecommunications industry in the United States, currently carried out under the US Telecom Act of 1996, provides one example. The Federal Communications Commission (FCC), the US regulator, faces strong opposition from the LECs over its implementation of the act. Most LECs are jointly or independently suing the FCC over various issues, including the FCC pricing rule and interconnection rule. According to these rules, LECs lease the unused portion of their network capacity to other telecommunications operators at reasonable rates (as defined by the FCC) and on a nondiscriminatory basis. The most prominent case is the battle in the Eighth Circuit Court of Appeals over the pricing rule, in which the LECs have so far prevailed against the FCC. Whatever the merits of this case, similar disputes are likely to punctuate the liberalization process in a number of countries.

The WTO Reference Paper, signed by 65 countries at the WTO Telecom Pact, was designed to curtail anticompetitive practices. Implementing this goal is a complex task because the telecommunications industry still reflects the legacy of natural monopoly. In chapter 5, International Trade in Telecommunications Service: An Economic Perspective, Henry Ergas explains how this legacy shaped the evolution of international trade in basic telecommunications services. Under the current system, the so-called cooperative model, telecommunications firms in different countries negotiate bilaterally to provide services, build facilities, and set technical and operational standards. Ergas argues that, although the cooperative model was born of necessity, it imposes significant economic costs today.

The most dominant negative aspect of the cooperation model is the gap between the price of international phone calls and the cost of providing the service. This gap is the result of three sets of charges: collection rates, settlement rates, and accounting rates. Telephone traffic is highly responsive to price; in economic jargon, the price elasticity of demand exceeds one. High collection rates—the rates that consumers pay to make international telephone calls—reduce the volume of outgoing calls and thereby diminish the revenue a telecom operator can collect from its domestic customers. At the same time, high collection rates increase the volume of incoming calls from overseas as firms and households initiate more calls from abroad, where collection rates are lower. National telephone operators, in turn, charge foreign telephone operators to terminate these incoming calls. The telecom-to-telecom termination fee is called the settlement rate, which is one half of the so-called accounting rate (the accounting rate is determined bilaterally between telecom operators). The revenue from foreign telephone operators increases as the volume of incoming calls increases, more than offsetting the decreased revenue from domestic customers. The combination of collection and accounting rates encourages monopoly telephone companies to set the domestic price of international phone calls quite high (well above costs).

This price distortion, however, generates huge consumer losses in a country with high collection rates. Callback and refile systems have emerged as a result. Although these systems increase consumer welfare by lowering the price of international telephone calls, they do not correct the fundamental price distortions. Ergas suggests that a strong international regulatory framework is necessary to get at these distortions.

On average, collection charges and accounting rates are higher in developing than in industrialized countries. In developing countries, international phone calls are an important source of revenue, which is needed to pay for the expansion of telecommunications networks. Thus, some argue that high collection and accounting rates simply reflect the need for capital to finance telecommunications investment. Charges and rates will tend to decrease as telecommunications networks improve. In this view, the price distortions are a temporary phenomenon, one which will be resolved over time.

Contrary to this rationale, Ergas demonstrates that accounting rates are determined by politics and regulation, not by the level of telecommunications development. According to Ergas's econometric findings, a country with an unstable political situation and a closed trade regime is likely to have high collection and accounting rates. This implies that cross-subsidies for network expansion are a small part of the story. In Ergas's view, without institutional changes, price distortions are likely to remain even after a telecommunications network is fully developed. Furthermore, Ergas emphasizes that a competitive environment in the telecommunications industry will not come about automatically.

In July 1997, the FCC proposed one solution to the issue of price distortion, an order to establish so-called benchmark rates. The order limits the settlement rates that foreign carriers may charge US telephone companies to complete international calls initiated in the United States. With guidance from the FCC, US telephone companies are negotiating with foreign carriers to reduce settlement rates to the benchmark levels over a five-year period. The proposed rates are determined by a country's income level: 15 cents per minute for upper-income countries, 19 cents for middle-income countries, and 23 cents for lower-income countries. The proposed benchmark rates are expected to slash the US deficit on trade in telecommunications services (now $5.4 billion annually). Many foreign telecommunications companies object to the FCC's order. The FCC, however, believes that they will eventually accept the benchmark rates, because the rates are well above the costs incurred by foreign carriers to complete international calls.

Improving ties between telecommunications companies in different countries, so that company-to-company charges are internalized, is viewed as another strategy for resolving the price-distortion issue. Robert W. Crandall analyzes mergers and joint ventures in chapter 6, Telecom Mergers and Joint Ventures in an Era of Liberalization. Advanced technology has expanded the opportunities for new telecommunications businesses, but it has also increased operating risks in the industry. Crandall predicts that, on balance, the new technology will lead to more competition in the telecommunications market. To ward off or prepare for the new competitive environment, traditional telephone companies are implementing three main strategies: attempting to protect their most profitable services, investing in newly liberalized regions, and entering new service areas such as video communication. Crandall believes that all three strategies encourage joint ventures and foreign investment. For instance, to establish a strong presence in foreign markets, many traditional telephone companies have recently merged or established joint ventures with foreign telecommunications firms and have increased direct investment in developing countries. Likewise, to offer complete video communications services, joint ventures between traditional telephone companies and cable, motion picture, and media firms are often the initial step.

Crandall questions whether mergers and joint ventures foster a competitive environment in the telecommunications market. Unfortunately, the answer may not always be yes. In his view, mergers and joint ventures are unlikely to lower high accounting rates to competitive levels, even though they help to solve the accounting rate problem for the joint-venture partners. Therefore, Crandall advocates a watchdog role for regulatory agencies such as the FCC to prohibit anticompetitive practices.

Crandall also encourages countries to allow foreign ownership. Foreign investment, particularly in developing countries, benefits the host country, because the country gains access to advanced technologies such as new fiber/cable broadband networks or direct satellite broadcasting without investing in telecommunications research.

Many countries fear the side effects of investment liberalization—job loss in the national telecommunications firm and foreign ownership of the essential communications system. Smaller countries are especially susceptible to this anxiety. Nonetheless, as Crandall explains, for smaller countries, building networks involves proportionately greater risks and raising capital is relatively more difficult. Thus, small OECD countries such as New Zealand and Australia have established liberalized telecommunications markets and have sold substantial stakes in their telecommunications firms to foreign operators. These actions suggest that many countries, even small countries, now believe that the benefits exceed the costs of liberalization.

Regardless of the size of the country, advanced telecommunications infrastructure helps economic growth through various channels. One of the channels is the financial sector. Jonathan D. Aronson focuses on the relationship between the telecommunications industry and the financial sector in chapter 7, Global Networks, Electronic Trading, and the Rise of Digital Cash: Implications for Policymakers. Reminiscent of the steel and auto industries, which grew in tandem, the telecommunications and financial service industries have mutually reinforced each other. Their affinity began when the flexible exchange rates system was introduced in the early 1970s. The rapid growth of the foreign-exchange market was partly prompted by better telecommunications, which reduced transaction costs and improved information flows. Individuals who have a computer, modem, and telephone line can now participate in cyberspace trading anywhere, anytime. However, as trading systems become automated, the possibility for system failure increases. In Aronson's view, this danger argues for a global surveillance of the telecommunications network.

In addition, new types of financial services are being invented through advanced telecommunications systems. Many credit cards are accepted worldwide, and their validity can be checked instantly via phone lines. Smart cards—a type of prepaid card—are replacing cash in Europe and Japan. These cards add convenience to our lives but require government surveillance to minimize system default and criminal activity. Lastly, advanced communications systems make it easier to access detailed information on individuals. This information is useful for analytical studies, but it also raises questions of personal privacy. Rather than depending on corporate morality to protect private information, Aronson suggests a global agreement on information usage.

The final part of this volume focuses on the status of telecommunications in selected APEC member countries. Chapter 8, Competition and Deregulation: An APEC Perspective, by Shin Cho and Myeongho Lee, examines the role of APEC in the liberalization of the telecommunications industry. The authors cite APEC leadership at the WTO Negotiations on Basic Telecommunications Services and the Information Technology Agreement (ITA) as a remarkable achievement. APEC nations pushed the ITA, partly because they appreciate the importance of the telecommunications industry in economic growth. In the same spirit, Association of Southeast Asian Nations (ASEAN) member countries, despite their traditionally conservative trade policies, offered remarkable promises at the WTO Negotiations on Basic Telecommunications Services.

Cho and Lee provide evidence on the link between economic growth and telecommunications infrastructure. They also demonstrate that, to establish an efficient telecommunications infrastructure, liberalization of the telecommunications market and encouragement of foreign investment should be the first priorities of policymakers. As already discussed, because building a telecommunications infrastructure demands capital and technology, it is more efficient for developing countries to invite foreign firms to develop the infrastructure than to develop it themselves. And to attract foreign firms or capital, state-owned telecommunications firms should be privatized. These steps are underway in most APEC member countries. Moreover, Singapore, Hong Kong, South Korea, and Japan are investing in neighboring Asian countries, using their financial resources and advanced technology.

Taken as a whole, the essays in this volume portray the current status of the telecommunications industry, give educated predictions, and lay out the following recommendations for maximizing the potential benefits from global liberalization:

- Competition must be vigorously implemented, country by country and market sector by market sector.

- Interconnection must be facilitated by harmonized technical standards and transparent protocols.

- Universal service arguments for monopoly telecommunications firms and high international accounting rates have little factual basis.

- Future regulatory attention should focus on surveillance of international joint ventures and mergers and on the security of the global telecommunications network.

II

TRADE AGREEMENT

2

Telecom Agreement Tops Expectations

JONATHAN D. ARONSON

On 15 February 1997, hours before their self-imposed deadline, trade and telecom negotiators successfully concluded the negotiations on basic telecommunications commitments that will supplement the services accords made in the 1994 Uruguay Round agreement. Sixty-nine countries whose telecommunications markets are worth $570 billion annually agreed to liberalize their telecommunications markets to varying degrees starting 1 January 1998. Many countries agreed to lower or remove domestic barriers to international competition in the provision of local, long-distance, and satellite services. Many also pledged to adopt transparent regulatory principles to support competition. The principle of technology neutral scheduling makes it possible for foreign and domestic service providers to furnish satellite services across borders on the same basis. Business and government leaders both expect that as competition rises, the rates for international calls will fall sharply.

Less than a year earlier, these same countries barely managed to extend negotiations and failed completely to find an agreement. What changed during that time to make agreement possible? What was agreed to? What will happen now?

Breakdown

As with most trade negotiations, the United States had the most to gain or lose, and was the *demandeur* for negotiations. Without US leadership

Jonathan Aronson is director of the School of International Relations and professor at the Annenberg School for Communications, University of Southern California. This chapter was originally published in Telecom International *1, no. 3 (May 1997): 17-20.*

and pressure, negotiations would never have started or would quickly have bogged down. The United States, arguing that all would benefit, offered to open its telecommunications markets to the world if others opened their markets in exchange. Yet, on 30 April 1996, Deputy US Trade Representative (USTR) Jeff Lang, the head of the US negotiating team, walked out of the telecom talks. The United States cited two specific problems: inadequate liberalization and regulatory safeguards for satellite services (especially global-mobile satellite services) and failure to agree to effective safeguards against anticompetitive behavior in the international telecommunications services market. Underlying both complaints was the failure to achieve a critical mass of countries to make significant offers. At the last moment the World Trade Organization (WTO) secretariat intervened and persuaded the participants to extend the negotiating deadline into early 1997.

Four Changes

More time alone would not have led to success. This chapter examines the unfolding of events from a distinctly US perspective. Obviously, this is skewed view of the events leading to the agreement; many other countries made major contributions as well. From the US standpoint, four shifts took place after the near collapse of April 1996. First, top officials in the United States—from President Clinton on down—focused on the telecom talks, decided to push for a bold agreement, and agreed not to accept a weak compromise.

Second, all the major WTO member countries were persuaded to make or improve their offers. Although Russia, China, and Taiwan were not at the table, the largest WTO member (in terms of international telephone traffic) not to make an offer was the United Arab Emirates. The 69 countries that made offers accounted for 99 percent of WTO telecommunications traffic. Third, the US satellite industry, much chastised as the cause of the April 1996 breakdown, went out of its way to work out its problems with the US negotiators. The industry was worried that the agreement might open the US market without guaranteeing it access. Fourth, the US industry was persuaded that the Federal Communications Commission's (FCCs) proposed new benchmark strategy, which is not part of the WTO agreement, would largely correct the problem of anticompetitive behavior.

US Leadership

The US negotiating team was committed, competent, and cohesive, working closely with industry, Congress, and each other from start to finish.

They had a vision of what they were trying to accomplish and worked hard to develop arguments and data that would persuade those at home and in other countries that an agreement was needed and would benefit all. US negotiators were able to draw on detailed FCC spreadsheets on the negotiations that helped back up their arguments about what various positions would yield. Moreover, with the election behind him, President Clinton took the unusual step of personally intervening with his counterparts at the November 1996 Asian Pacific Economic Cooperation (APEC) meeting of heads of state at Subic Bay, Philippines.

More and Better Offers

In April 1996 the United States was troubled that many crucial countries, particularly in Asia, had not made offers. President Clinton's efforts at the APEC meeting convinced others that the United States was serious and helped persuade leaders in Malaysia, Indonesia, and elsewhere in Asia to make or improve offers. The United States employed a novel tactic at the November trade ministers meeting in Singapore. Acting USTR Charlene Barshefsky and FCC Chairman Reed Hundt convened a special session of senior trade and telecom officials designed to pressure Asian countries to bring forward serious offers. The White House was involved as needed. In early February, the president and the cabinet decided to go for all or nothing. The decision was taken to achieve a major breakthrough, not to weaken the US offer and accept a weak compromise.

The Satellite Issue

Not surprisingly, in the early phases of talks the US negotiators spent much more time with AT&T, MCI, and the other telephone companies than with the satellite industry. The negotiators did not fully analyze how the agreement might affect US satellite firms. Although skilled in the ways of the International Telecommunications Union (ITU), the satellite industry did not recognize how powerful and sweeping a WTO obligation could be. Even those satellite officials attending USTR briefings did not fully grasp the point. When they did finally understand, the industry quickly dissolved into sharply divided factions between those involved in new global mobile technologies and those that were not. Most problems ultimately proved less severe than imagined, but there was no time to work through the issues before the April 1996 deadline.

Some contended that the global mobile satellite systems killed the April deal. In fact, opposition in the satellite industry was far more widespread, but it is instructive to consider how those systems affected negotiations. The three main US satellite firms licensed by the FCC—Iridium, backed by Motorola; Globalstar, backed by Loral; and Odyssey, backed by TRW and Canada's Teleglobe—became worried that the agreement might allow International Communications Organization (ICO), the fourth entry, not only to enter the US market but also to use its close relationship with Inmarsat and the dominant operators in some developing countries to hamper the access of the three other satellite service providers around the world.

At the October 1996 ITU World Policy Forum, industry and the US government worked cooperatively to establish principles that would enhance the probability that all global mobile systems would be treated equally and national regulations would evolve to remove technical obstacles to the success of the systems. WTO discussions also led to the idea of technology-neutral scheduling and demonstrated that the regulatory principles substantially decreased other competitive risks.

Benchmark Pricing

At the time of the April 1996 breakdown, US companies and negotiators worried that many countries would not open their markets for international telephone services and would not commit to the simple resale of capacity. The United States was concerned that a WTO agreement might allow carriers from closed foreign markets to enter the US market freely and use resale to provide international telephone service at roughly 7 cents a minute. But US carriers would have to rely on the archaic and bloated cost structure that the international accounting rate and settlement system specified. The average cost a minute for US carriers was about 40 cents under this system. Before the talks broke down in April, US negotiators tried to persuade other countries that the United States should be able to deny a license to foreign carriers if their entry created a competition problem in international services. Other countries did not agree. By February 1997, the FCC proposed a unilateral regulatory initiative to place price caps on US accounting rates to all countries that would be phased in over several years. The Benchmark Order issued by the FCC in August 1997 will require immediate compliance with the price cap for any foreign carrier seeking permission to serve its home market from the United States after January 1998. In that way, the FCC can prevent, for example, France Telecom from securing an accounting rate advantage over AT&T on the same route (recall that France Telecom settles accounts internally, but AT&T must settle with France Telecom) when carrying calls from New York to Paris.

End Game

In the weeks leading up to the 15 February 1997 talks, US negotiators believed they had about a 50-50 chance of success. Pressures came from many directions. At home, skeptics in Congress, led by Senator Fritz Hollings (D-SC), doubted that the United States would gain enough to justify opening its markets to the world and even questioned the authority of negotiators to make the offer to permit up to 100 percent foreign ownership. Senator Hollings reminded US negotiators that Acting USTR Charlene Barshefsky still required a waiver before she could be approved because she had once represented Canada. (She was approved overwhelmingly on 4 March.) This Congressional pressure brought home to negotiators that the United States needed to address seriously the issue of restrictions on foreign investment rights in critical countries, especially its NAFTA partners Canada and Mexico. Industry still was nervous about the final deal, and doubt remained that a sufficient mass of acceptable offers could be brought forward and held together.

At the final negotiating sessions in Geneva, new and better offers materialized. The 12-person US negotiating team, led by Deputy USTR Jeff Lang, was surrounded and often cheered on in Geneva by Congressional and business observers and was strongly supported in Washington. Barshefsky and Hundt worked the phones in Washington to make certain that problems were solved, commitments were met, and offers were not withdrawn.

France's insistence that video services be excluded threatened to pull apart the EU position. The European Union revised its offer to allow for the French objections, but it also made it more difficult for France and others to exclude specific telecommunications services. Except for certain broadcast-like services sent over the Internet, efforts by France or others to find exceptions to the agreement can be challenged.

When Argentina pulled its offer on 14 February because the United States was instituting WTO actions involving trade disputes in other areas against it, the US negotiators worried that other countries might follow suit and the negotiations could collapse. Senior White House officials intervened directly with their Argentine counterparts and persuaded them to put an offer back on the table. When a negotiator from another country started to freelance, offering less than his government had promised, the officials responsible in the capital were alerted and they intervened to set things straight.

The United States was particularly troubled that Canada, citing cultural issues, continued to insist on significant restrictions on foreign investment in the Canadian industry. The Canadians' offer restricted foreign investment in facilities-based suppliers to no more than 46.7 percent. Direct stakes were limited to 20 percent. Restrictions on satellite services and severe routing restrictions designed to protect Teleglobe for the next several years also irritated US firms and negotiators. Once Mexico agreed to permit 49 percent foreign investment across the board and 100 percent foreign ownership of cellular systems, it provided a far stronger offer than Canada's.

In the end, the United States decided to make no market access or national treatment commitments and to take a most favored nation (MFN) exemption for direct-to-home television and for digital-audio satellite services. This exemption allowed the United States complete discretion in choosing which countries could provide direct broadcast services to the US market, a matter of significant importance to Canada.

Other countries were important contributors to the eventual success of the negotiations at critical moments. Without a strong EU offer, for example, it would have been impossible to attract others to the table. In Latin America, Peru showed considerable courage and creativity early in the negotiations and Brazil demonstrated high resolve and ingenuity in the midst of domestic reform by tabling a credible offer. In Asia, Hong Kong aided the work on regulatory principles early in the negotiations, and both Japan and South Korea were more flexible than in the past on the same subject. After the April failure, Singapore decided to table a strong offer and actively push other Association of Southeast Asian Nations (ASEAN) countries to do likewise. These efforts led to breakthroughs in countries like Malaysia, Indonesia, the Philippines, and Pakistan.

Results

The negotiations yielded market access, adoption of regulatory principles, liberalization of foreign investment rules, and satellite offers. In the US coding system, which is adopted here, a full offer does not depend on the date on which a country comes into compliance.

Market Access

According to US government figures, before the agreement 17.1 percent of the markets of the top 20 US trading partners were competitive. When the WTO agreement is fully implemented, 97.2 percent of those same markets will be open to at least some competition. Between April 1996 and February 1997, 21 countries—including Indonesia, Malaysia, Jamaica, and South Africa—made new offers. During the same period, 25 countries and the 15-strong European Union improved their offers. In April 1996, there were 48 market-access agreements on the table, only 33 of which could classify as fully opening. By the conclusion of negotiations, 52 commitments to provide full market access for all services and facilities had been made. An additional 5 offers provided for selected service openings, and 12 countries made limited or no market-access commitments. Countries making access offers accounted for 97 percent of all international minutes of traffic of WTO member markets.

Regulatory Principles

Between April 1996 and February 1997, the number of countries agreeing to adopt transparent, pro-competitive regulatory principles as defined by the WTO rose from 33 to 59. Six countries—Bolivia, India, Morocco, Pakistan, the Philippines, and Turkey—adopted some of the reference paper or separate regulatory principles. Ecuador and Tunisia were the most important countries that failed to adopt regulatory principles. Countries that did adopt the principles represented 99 percent of the WTO telecommunications market and 97 percent of the minutes of international traffic in 1995.

Foreign Investment

In April 1996, there were 24 offers to permit foreign ownership in excess of 50 percent for all services and facilities effective on 1 January 1998 or phased in later. An additional 24 countries offered to permit significant foreign investment in their markets, either allowing controlling investment for limited services or limiting stakes to less than 50 percent for all services. When the negotiations concluded, 44 full offers were on the table, and another 25 countries had made significant offers on foreign investment. Bangladesh, Brazil, Canada, Ivory Coast, Ecuador, Ghana, Hong Kong, Israel, South Korea, Mexico, Poland, and Tunisia offered to permit control of some, but not all, services. Dominica, India, Indonesia, Malaysia, the Philippines, Senegal, South Africa, Sri Lanka, Thailand, and Turkey agreed to permit some foreign investment, but not control, to outsiders.

Satellite Services

In April 1996, there were 28 full and 7 selected satellite offers on the table. By February, these numbers increased to 50 full offers guaranteeing market access for all domestic and international satellite services and facilities. Six additional countries—Brazil, Colombia, Ivory Coast, Ghana, Hong Kong, and South Africa—guaranteed market access for selected satellite services and facilities. Thirteen of the 69 parties to the WTO agreement made no market-access commitments for satellite services or facilities. These were Bangladesh, Belize, Dominica, Ecuador, India, Morocco, Pakistan, Philippines, Papua New Guinea, Senegal, Sri Lanka, and Tunisia.

Table 2.1 shows that 23 of the 25 European and Eastern European countries made full offers in all four categories. However, the offers of all 6 Eastern European participants, Turkey, Greece, Ireland, and

Table 2.1 Offers made at the February 1997 WTO telecom negotiations

Country	Offer after 30 April 1996	Market access offer	Regulatory principle offer	Foreign investment offer	Satellite offer
Europe					
Austria	I	F-1998	F	F-1998	F-1998
Belgium	I	F-1998	F	F-1998	F-1998
Denmark	I	F-1998	F	F-1998	F-1998
Finland	I	F-1998	F	F-1998	F-1998
France	I	F-1998	F	F-1998*	F-1998
Germany	I	F-1998	F	F-1998	F-1998
Greece	I	F-2003	F	F-2003	F-2003
Iceland	I	F-1998	F	F-1998	F-1998
Ireland	I	F-2000	F	F-2000	F-2000
Italy	I	F-1998	F	F-1998ᵇ	F-1998
Luxembourg	I	F-1998	F	F-1998	F-1998
Netherlands	I	F-1998	F	F-1998	F-1998
Norway	I	F-1998	F	F-1998	F-1998
Portugal	I	F-2000	F	F-2000ᶜ	F-2000
Spain	I	F-1998ᵈ	F	F-1998*	F-1998ᵈ
Sweden	I	F-1998	F	F-1998	F-1998
Switzerland	I	F-1998	F	F-1998	F-1998
Turkey		F-2006	P	P-49%	F-2006
United Kingdom	I	F-1998	F	F-1998	F-1998
Eastern Europe					
Bulgaria	N	F-2005	F	F-2005	F-2005
Czech Republic	I	F-2001	F	F-2001	F-2001
Hungary		F-2004	F	F-2004	F-2004
Poland	I	F-2003	F	Sᶠ	F-2003
Romania	N	F-2003	F	F-2003	F-2003
Slovak Republic	I	F-2003	F	F-2003	F-2003

Notes: Dates indicate the year in which the schedule for the country takes effect, generally on 1 January. Percentage in the investment column indicates the percentage of foreign ownership allowed. C = Commitment to adopt procompetitive regulatory principles in the future. F = Full offer. I = Improved offer made since 30 April 1996. N = New offer made since 30 April 1996. P = Partial commitments. S = Selected commitments. 0 = No offer or commitment.

* These countries maintain foreign investment limits below control.
a. Except France Telecom.
b. Except Stet.
c. Will review limit.
d. Starting 1 December 1998.
e. Except Telefonica.
f. 100 percent for local wireline (voice and data); 49 percent for wireless, international and long distance voice and data.

(table and notes continued on next page)

Table 2.1 Offers made at the February 1997 WTO telecom negotiations (continued)

Country	Offer after 30 April 1996	Market access offer	Regulatory principle offer	Foreign investment offer	Satellite offer
North America, Caribbean, and Central America					
Antigua and Barbuda	N	F-2012	0	F-2012	F-1998
Belize	N	0	0	0	0
Canada	I	F-1998*	F	S^g	F-2000^h
Dominica	N	0	F	P^i	0
Dominican Republic	I	F-1998	F	F-1998	F-1998
El Salvador	N	F-1998	F	F-1998	F-1998
Grenada	N	F-2007	F	F-2007	F-2007
Guatemala	N	F-1998	F	F-1998	F-1998
Jamaica	N	F-2013	F	F-2013	F-2013
Mexico	I	F-1998	F	S^i	F-2002
Trinidad & Tobago		F-2010	F	F-2010	F-2010
United States	I	F-1998	F	F-1998	F-1998
South America					
Argentina		F-2000	F	F-2000	F-2000
Bolivia	N	F-2001	0	F-2001	F-2001
Brazil	I	S**^{k}	C	S^l	S^m
Chile		F-1998	F	F-1998^n	F-1998
Colombia	I	0°	F	F-1998	S^p
Ecuador		0	0	S^q	0
Peru	I	F-1999	F	F-1999	F-1999
Venezuela		F-2000	F	F-2000	F-2000

g. 100 percent for resellers and mobile satellite service providers; 100 percent for fixed satellite services as of March 2000; 100 percent for submarine cable licenses as of 1 October 1998; 46.7 percent on all other services.

h. 2000 for fixed, 1998 for mobile.

i. Unbound.

j. 100 percent for cellular services; 49 percent for all other services and facilities.

k. Open for all nonpublic services for closed user groups not connected to public switched network but will bind legislation expected to cover public and nonpublic. Must route through Brazilian gateway.

l. 100 percent for nonpublic services; 100 percent for cellular and satellite services after 20 July 1999.

m. Open for all nonpublic domestic and international services for closed user groups (not connected to public switched network). Will bind future reform legislation, which is expected to cover all services within one year of enactment. Requirement to route all international traffic through gateway in Brazil. Brazilian or foreign satellites can be used but preference given for Brazilian satellite supply when they offer better service or equivalent conditions. No foreign ownership restrictions after 20 July 1999.

n. Except local service.

o. Subject to economics needs test.

p. Open for geostationary satellites.

q. 100 percent for cellular only.

(table and notes continued on next page)

Table 2.1 Offers made at the February 1997 WTO telecom negotiations (continued)

Country	Offer after 30 April 1996	Market access offer	Regulatory principle offer	Foreign investment offer	Satellite offer
Asia					
Australia	I	F-1998	F	F-1998^r	F-1998
Bangladesh	N	0	C	S^s	0
Brunei	N	F-2010	F	F-2010	F-2010
Hong Kong		S^t	F	S^u	S^v
India	I	0^w	0	P-25%	0
Indonesia	N	0	F	P-35%	0
Japan		F-1998	F	F-1998^x	F-1998
South Korea		F-1998*	F	S^y	F-1998
Malaysia	N/I	F-1998*^{z}	F	P-30%	F-1998
New Zealand	I	F-1998	F	F-1998^{aa}	F-1998
Pakistan	I	S^{bb}	0	F-2004	0
Papua New Guinea	N	0	F	0	0
Philippines	I	F-1998*	0	P-40%	0
Singapore	I	F-2000	F	F-2000	F-2000
Sri Lanka	N	0^{cc}	F	P-40%	0
Thailand	I	F-2006*	C	P-20%	F-2006
Africa and Middle East					
Ivory Coast	S^{dd}	F	S^{ee}	S^{ff}	
Ghana	N	S^{gg}	F	S^{hh}	S^{ii}
Israel	I	F-2002	F	S^{ii}	F-2002
Mauritius	I	F-2004	C	F-2004	F-2004
Morocco	I	0	0	P^j	0
Senegal	N	F-2006	F	P-35%	0
South Africa	N	0^{cc}	F	P-30%	S^{kk}
Tunisia	N	0	0	S^{ll}	

r. Except Vodaphone and Telestra.

s. 100 percent for cellular and two additional local and long distance carriers.

t. Resale of voice, data, and fax; call back and closed user groups.

u. 100 percent for resale of voice, data, fax; closed user groups.

v. Open for all mobile satellite services and self provision of external satellite circuits by a company or closed user group. Interconnection to public switched network at Hong Kong end not permitted.

w. Subject to review in 2004.

x. Except KDD and NTT.

y. 49 percent for resale until 1 January 2001, thereafter 100 percent; 33 percent for facility providers until 2001, thereafter 49 percent, with individual shareholding limited to 33 percent for wireless and 10 percent for wireline; 20 percent limit in Korea Telecom until 1 January 2001, thereafter 33 percent with individual shareholding limited to 3 percent.

z. Through existing suppliers.

aa. Except 49.9 percent limit in Telecom NZ for any one single foreign entity.

(notes continued on next page)

**Table 2.1 Offers made at the February 1997 WTO telecom
negotiations** (continued)

bb. Telex and fax.

cc. Duopoly.

dd. Analog cellular, mobile, PCS, and nonvoice satellite-based services, data transmission and private leased circuit services.

ee. 100% for all services other than point-to-point telephony and telex.

ff. Open for all services except fixed voice and telex.

gg. Closed user groups.

hh. Only in joint venture with Ghanaian nationals.

ii. Open for global mobile; open for domestic fixed (excluding public voice) through joint ventures with Ghanaian nationals.

jj. 80 percent for cellular; 74 percent for international services; 100 percent for value-added services.

kk. Will schedule unspecified commitments within one year of adoption of legislation.

ll. Telex and data transmission in 1999; mobile, paging, and teleconferencing in 2000; and local in 2003.

Sources: Compiled from data provided.

Portugal from the European Union will not take effect until at least two years after the others. Poland was also less forthcoming on foreign investment for wireless and international long distance for voice and data, and Turkey offered only 49 percent control of foreign investment and was weak on regulatory principles.

Eleven of the 20 offers from the Americas were strong across the board, but there were some surprises. The Dominican Republic, El Salvador, Guatemala, United States, and Chile's offers all take effect on 1 January 1998. Offers by Argentina, Bolivia, Peru, and Venezuela and the weaker offer by Colombia will be in place by 2001. At the other extreme, the offers of Belize, Dominica, and Ecuador are extremely weak, and offers by Grenada, Antigua and Barbuda, Trinidad and Tobago, and Jamaica do not take effect for a decade or more. Canada, Mexico, and Brazil all maintained restrictions of varying severity on foreign investment.

Of the 16 Asian countries, only Australia, Japan, New Zealand, and Singapore will fully liberalize their markets in all four categories on schedule in 1998, and all of these except Singapore will continue to maintain restrictions on foreign investments in their major carriers.

The offers from Papua New Guinea and Indonesia are disappointing, and all four of the South Asian offers (Bangladesh, India, Pakistan, and Sri Lanka) were quite weak, with India's by far the weakest. The other Asian offers were strong in some areas but weak in others.

None of the eight African and Middle Eastern countries made fully opening offers in all four categories. Israel and Mauritius moved the furthest of this group.

Next Steps

The WTO negotiations exceeded all expectations. Industry officials from the United States and around the world enthusiastically endorsed the results.

Still, there is much to be done to turn this agreement into reality. Many countries, including the United States, will need to bring regulations into conformity with their new WTO obligations. The United States, for example, will have to change the Effective Competitive Opportunities test it now uses when considering granting access to foreign firms wishing to acquire more than 25 percent of a US firm. The big question is how countries will use the WTO dispute settlement mechanism and the application of the regulatory principles to speed the growth of competition and the reduction of international telephone rates.

3

Assessing the WTO Agreement on Basic Telecommunications

WILLIAM J. DRAKE AND ELI M. NOAM

On 15 February 1997, the Group on Basic Telecommunications (GBT), organized under the auspices of the World Trade Organization (WTO), completed negotiations on the world's first multilateral deal liberalizing international trade in basic telecommunications services. The deal that was struck comprises 55 schedules, covering 69 governments (the European Commission negotiates on behalf of European Union member governments and submitted one schedule for all of them, hence the numerical discrepancy), of legally binding commitments to open some or all of the participating countries' basic telecommunications markets to foreign competition.[1]

The successful conclusion of the basic telecommunications negotiations, which lasted from 1994 to 1997, was greeted with great enthusiasm by the

William J. Drake is associate director of the Communication, Culture and Technology Program at Georgetown University, Washington. Eli M. Noam is professor of finance and economics and director of the Columbia Institute for Tele-Information at Columbia University, New York. A shorter version of this paper appears in the journal Telecommunications Policy, *November 1997.*

1. The 55 schedules (covering 69 governments) annexed to the Fourth Protocol of the General Agreement on Trade in Services (GATS) covered the following countries: Antigua and Barbados, Argentina, Australia, Bangladesh, Belize, Bolivia, Brazil, Brunei Darussalam, Bulgaria, Canada, Chile, Colombia, Cote d'Ivoire, Czech Republic, Dominica, Dominican Republic, Ecuador, El Salvador, European Communities and its Member States, Ghana, Grenada, Guatemala, Hong Kong, Hungary, Iceland, India, Indonesia, Israel, Jamaica, Japan, South Korea, Malaysia, Mauritius, Mexico, Morocco, New Zealand, Norway, Pakistan, Papua New Guinea, Peru, Philippines, Poland, Romania, Senegal, Singapore, Sri Lanka, Switzerland, Slovak Republic, South Africa, Thailand, Trinidad and Tobago, Tunisia, Turkey, United States, and Venezuela. Two other countries—St. Vincent and the Grenadines, and the Bahamas—also devised schedules, but were unable to do so by 15 February 1997; hence, these were not attached to the protocol.

WTO secretariat, the international trade policy community, and many governments and private firms. Washington in particular was full of undisguised elation as top policymakers hit the conference circuit and talked up the agreement to the press and the Congress (Barshefsky and Hundt 1997). President Bill Clinton weighed in by issuing a press release stating that "the American-led negotiations in Geneva have resulted in a landmark agreement . . . expected to grow [the world market] to more than $1 trillion over the next 10 years. . . . Today's agreement will bring clear benefits to American workers, businesses, and consumers alike— new jobs, new markets and lower prices—and will spread the benefits of a technology revolution to citizens around the world." In parallel, major corporations that stood to benefit directly from the deal immediately released a slew of congratulatory pronouncements, while those that are less certain to benefit from the prospect of enhanced foreign competition in their markets went along with the party line, albeit with more muted praise.

The exuberance is certainly understandable, especially on the part of the negotiators who spent almost three painstaking years running back and forth to the WTO and a multiplicity of bilateral meetings in an effort to reach an agreement. But is the elation justified? Will the GBT deal unleash a cosmic "big bang" of competition and dynamism in national and international markets, as its proponents suggest? In the United States at least, the affirmative answer seems to be an article of faith among many observers, even though there has been precious little public debate about precisely what the deal entails. Indeed, the GBT process (like the WTO more generally) was not quite a model of open and transparent decision making, so a full understanding of what it produced may not be widespread beyond the circle of international trade policy insiders.

With these considerations in mind, this chapter assesses the GBT deal. Our objective is not to analyze the details of the negotiation process and explain the precise shape of the outcome in terms of the bargaining dynamics and who won or lost on particular issues; that is for another time. Instead, our goal is more modest—to consider, in the absence of real public debate on the matter, whether or not the deal is as significant as is widely assumed. To do so, the chapter adopts the somewhat unusual strategy of bringing together two authors with divergent views on the issue.

Sections one and two were written by William J. Drake, who argues that the GBT deal could, depending on its implementation, have a substantial liberalizing effect not only on specific markets but also on the broader institutional arrangements of the global telecommunications policy environment. Section three was written by Eli M. Noam, who argues that the GBT is useful, but that its significance is being greatly exaggerated, that most policy changes were taking place anyway, and that it could in some cases have the negative effect of slowing down the process of global liberalization. On neither score are the arguments presented intended to be exhaustive; rather, they merely are intended to be suggestive and to help to stimulate further discussion.

The GATS Framework and International Telecommunications

When the GBT accord was announced, the WTO and US government agencies received a flurry of calls from the press and other interested parties for copies of the text. Apparently, the callers were making the reasonable assumption that the GBT deal is embodied in a new international regime treaty with its own elaborate set of provisions. But this is not the case. Instead, the deal comprises additions to 55 national schedules of market opening commitments on international trade in commercial (i.e., nongovernment) services that were made in accordance with an international regime established in 1994—the General Agreement on Trade in Services (GATS). Hence, in order to understand how the GBT deal works, one must understand the form and operation of the GATS regime.

The establishment of the GATS was one of the key results of Uruguay Round trade negotiations that were launched at Punta del Este in September 1986 and concluded in December 1993. The agreements formally signed at a ministerial meeting at Marrakesh in April 1994 comprised a series of interrelated negotiations among the 125 member governments of what was then the General Agreement on Tariffs and Trade (GATT). These produced four major outcomes. First, the GATT's intergovernmental organization was replaced by the WTO—which, among other things, has a broadened mandate and a strengthened dispute resolution system —while the GATT's multilateral trade rules and related arrangements became part of a larger set of agreements administered by the WTO.

Second, the GATT regime for trade in goods was updated and expanded to address issues like trade-related investment measures and nontariff barriers (NTBs) on agricultural products. Third, a new regime on trade-related aspects of intellectual property rights (TRIPs) was established. And fourth, the GATS regime was created. The new organizational structure of the WTO reflects these outcomes; the three regimes are administered by corresponding bodies—the Council for TRIPs, the Council for Trade in Goods, and the Council for Trade in Services— each of which is broken down into a series of subsidiary groups focusing on specific issues or sectors.[2]

---

2. See WTO (1994). Under the Council for Trade in Services, there will be groups focusing on trade in telecommunications services, financial services, professional services, maritime services, the movement of natural persons across borders to deliver services, and perhaps other issues.

Ever since the 1970s, when analysts began to think about international transactions in services as a trade issue, most observers believed that telecommunications would be central to any sort of future multilateral regime on trade in services.[3] This is because telecommunications services play a dual role in the larger realm of international transactions in the service sector. First, as a dynamic and lucrative industry in its own right, it attracted the attention of policymakers, analysts, and private sector proponents of services liberalization. Today's global telecommunications services markets have been estimated to be worth about $600 billion per year, $100 billion of which is said to involve cross-border trade.[4]

Second, telecommunications is a key means of moving other types of disembodied (that is, temporally and spatially separated from the producer) information-intensive services—e.g., financial, management consulting, audiovisual, and advertising services—across national borders. Uruguay Round participants recognized that, as an infrastructure for the delivery of other services, telecommunications networks and services therefore constituted a crucial form of cross-border supply. Cross-border supply (via telecommunications, postal systems and so forth) is one of four designated modes of supply for services under the GATS regime, the others being movement of the consumer to the producer's country, movement of a natural person producer to the consumer's country, and the commercial presence of producer firms in consumer countries, notably via foreign direct investment (FDI).

Hence, telecommunications as a cross-border mode of supply was an important issue in the GATS negotiations. Any agreement to liberalize international trade in other services that depend on telecommunications (e.g., banking, professional services, and so on) had to allow their providers a right to gain access to and use public switched networks and services under reasonable conditions, just as providers of goods need reasonable access to roads, waterways, airports, and ports. That meant that the GATS regime had to contain special provisions dealing with the unique status of telecommunications as a cross-border mode of supply.

The GATS regime established by the Uruguay Round is made up of three major components. The first is a Framework Agreement, which includes the overarching principles of liberalization—the General Obligations and Disciplines (GODs). Some of these principles were adopted

---

3. On the evolution of trade in services thinking before and during the Uruguay Round and its impact on the GATS negotiations, see Drake and Nicolaïdis (1992, 37-100). On the nature of the GATS agreement and its implications for the networked global information economy, see Nicolaïdis (1995, 269-302).

4. These figures are from ITU (1997). They reflect the fact that "jointly provided" telecommunications services are traded, in that countries whose carriers pay foreign carriers to terminate their outbound traffic are importing a service, while the countries of the terminating carriers are exporting a service.

from the traditional GATT regime for trade in goods, with necessary modifications to take into account the differences between goods and services trade. Others were created to address specific issues and some NTBs that do not apply to trade in goods.

There are 15 GODs plus three corollary principles: most favored nation treatment (MFN), meaning nondiscrimination between WTO members; transparency; disclosure of confidential information; increasing participation of developing countries; economic integration agreements and a corollary on labor market integration; domestic regulation; mutual recognition of standards or criteria for the authorization, certification, and licensing of service suppliers; monopolies and exclusive service providers; business practices; emergency safeguard measures; payments and transfers; restrictions to safeguard the balance of payments; government procurement; general exceptions; national security exceptions; and subsidies. Some of the GODs are binding on states across-the-board—i.e., for all measures affecting international trade in services in the absence of specific provisions to the contrary—while others apply only when states make liberalizing commitments. Either way, all of them can apply to the basic telecommunications concessions unless a government has taken an exception or waiver.

In addition to the GODs, the Framework Agreement includes Specific Commitments, which are negotiated undertakings by each government to liberalize the provision of particular service sectors or subsectors according to the four modes of supply. In other words, a given type of service transaction is opened to competitive supply only insofar as a government agrees to do so; it can pick and choose what to liberalize and what not to liberalize. Thus, for example, a government can choose to allow the cross-border supply of a particular service while not allowing supply via commercial presence, or vice versa.

There are three kinds of Specific Commitments. GATS Article XVI pertains to market access, a term that is defined as the removal of six specific kinds of NTBs on each of the four modes of delivery. These include: (a) limitations on the number of service suppliers, whether in the form of numerical quotas, monopolies, exclusive service suppliers, or economic needs tests; (b) limitations on the total value of service transactions or assets in the form of numerical quotas or needs tests; (c) limitations on the number of service operators or the quantity of service outputs; (d) limitations on the total number of natural persons that may be employed in a particular service sector; (e) measures that restrict or require specific types of legal entities or joint ventures through which a supplier may provide a service; and (f) limitations on the participation of foreign capital expressed in terms of the maximum percentage of foreign share holdings or the total value of investment (WTO 1994, 342). Hence, market access commitments involve reducing or eliminating quantitative restrictions on foreign participation in a given market.

GATS Article XVII concerns national treatment, or the obligation to accord foreign services and service suppliers treatment no less favorable than what a country affords its own services and suppliers. This is important in telecommunications, but in truth most governments have not historically had formalized derogations from the national treatment standard on which concessions had to be listed in the GBT. Finally, GATS Article XVIII allows members to undertake additional commitments with regard to services measures not covered by market access and national treatment, including commitments on qualifications, standards, and licensing.

The second major component of the GATS is the annexes. These clarify or modify how the GODs pertain to some issues specific to particular service sectors and modes of supply, and they also establish the legal basis for future negotiations in certain sectors. There are eight annexes, dealing with Article II MFN exemptions, the movement of natural persons supplying services, air transportation services, financial services (two annexes), negotiations on maritime transport services, telecommunications, and negotiations on basic telecommunications. These last two are of particular interest here.

The Annex on Telecommunications establishes obligations for governments to ensure access to and use of public telecommunications transport networks and services. In other words, public telecommunications systems must be accessible to foreign service providers for domestic and international service provisioning once governments have agreed to liberalize the service in question, be it financial services, professional services, or whatever. The annex requires that access to and use of public networks and services be provided on a reasonable and nondiscriminatory basis. Moreover, governments must ensure that foreign suppliers also have access to and use of private leased circuits, and that they can: (1) purchase or lease and attach terminal or other equipment interfacing with public networks; (2) interconnect private leased or owned circuits with public networks, or with circuits leased or owned by another service supplier; and (3) use operating protocols of the service supplier's choice, provided they do not disrupt the telecommunications transport networks and services to the public generally. In addition, governments must ensure that foreign service suppliers can use these networks and circuits to transfer information without undue impediments within and across national borders, and that they can access information contained in databases held in any member country. This last provision is an echo of the transborder data flow debate of the 1970s and 1980s, when many companies worried that governments would, for protectionist purposes, limit their ability to send and gain access to computerized information over transnational networks.[5]

---

5. The evolution and resolution of the transborder data flow debate—which was closely related to the move toward trade treatment for telecommunications and information services—is assessed in Drake (1993, 259-313).

The Annex on Telecommunications requires that governments apply no conditions on access and use other than as necessary to safeguard the public service responsibilities of public telecommunications operators (PTOs), protect the technical integrity of public networks, or ensure that foreign service suppliers only provide services that have been designated open to competition. However, if they meet these criteria, governments may adopt policies that: (1) restrict resale and shared use of public services; (2) require the use of specific technical interfaces and protocols for interconnection; (3) require the interoperability of services; (4) require type approval of terminal or other equipment interfacing with public networks; (5) restrict the interconnection of private leased or owned circuits with either public networks or the circuits of other service suppliers, and (6) require the registration or licensing of foreign suppliers (WTO 1994, 359-363).[6] Clearly, these are potentially rather restrictive measures that a government could use to impede a service supplier's access to and use of public networks and leased circuits. How easily one might get away with doing this would depend on bilateral consultations over their necessity to achieve the three goals, and if such consultations fail, the results of the WTO dispute settlement process.

The other annex of relevance here concerns negotiations on basic telecommunications. During the Uruguay Round, 60 governments submitted 48 schedules (one for the then 12 members of the European Union) of Specific Commitments on telecommunications as a trade sector in its own right (as opposed to a mode of supply, which is covered by the annex discussed above). The process of agreeing to such commitments, under the umbrella responsibilities of the GODs, was a politically and intellectually difficult task that consumed a great deal of energy and bargaining time. In general, the 48 schedules included commitments on computer-enhanced or value-added services. While 19 schedules included some commitments on basic telecommunications, usually these were very narrowly drawn, for example, liberalizing the provision of discrete services like mobile paging or home country direct calling via select modes of supply. Moreover, most of these rather limited commitments were made by developing countries, Japan being the primary exception.

Hence, there was widespread recognition that cutting a broader deal on basic telecommunications would require additional negotiations. The Annex on Negotiations on Basic Telecommunications provided a legal basis for launching these negotiations with a view toward incorporating more and more significant basic telecommunications concessions into the GATS framework. The annex provided that Article II MFN treatment and exceptions would enter into force for such commitments only upon

---

6. For analyses of the Annex on Telecommunications and related GATS instruments, see Tuthill (1996, 89-99); Pipe (1993); and ITU (1996).

the completion of the post-Uruguay Round negotiations or, should they fail, only when negotiators issued their final report.

Finally, the third major component of the GATS is the National Schedules, in which governments inscribe their Specific Commitments on market access, national treatment, and additional commitments for each service sector or subsector according to the four modes of supply. Comprising several thousand pages, the schedules include a wide range of market opening measures in different service sectors, including telecommunications. The recent deal on basic telecommunications, then, simply involved folding new commitments into the schedules under the legal framework of the GATS, rather than establishing a new treaty.

By way of summary, the structure of the GATS regime is depicted in box 3.1. In addition, to depict how the governments go about listing Specific Commitments on services according to the four modes of delivery, table 3.1 presents a model schedule for basic telecommunications concessions that was prepared by the WTO secretariat.

In accordance with the GATS annexes, a Negotiating Group on Basic Telecommunications (NGBT) was launched in April 1994. At the outset only 33 governments agreed to participate, but over time more governments joined the process as participants or observers. Pending conclusion of the negotiations, participants agreed not to apply any measures affecting trade in basic services that would affect their bargaining positions and leverage, and set April 1996 as a date by which the negotiations would conclude. Beginning in May 1994, 15 meetings were held at the WTO headquarters in Geneva. Members attempted to clarify the current situation with respect to national regulatory regimes, explore the complex issues involved, and move toward consensus on what kinds of services should be included in the schedules in what format. In addition, many so-called rump groups were created to discuss matters outside the formal institutional structures of the WTO, and intensive bilateral negotiations were held among participating governments in an effort to secure concessions of particular interest to their service suppliers.

The NGBT confronted difficult conceptual issues. For example, Australia proposed that to deal with the market distortions caused by high international accounting rates, so-called termination services should be scheduled as market access limitations. This idea was not endorsed by other governments; instead, they concluded that enhanced competition resulting from the agreement would reduce accounting rates over time. Other issues included whether national schedules should enumerate nondiscriminatory limitations on the number of service suppliers in situations where numerical restrictions are imposed for strictly technical reasons like spectrum scarcity; how to regard universal service and public interest obligations that might have trade restricting effects, even though such effects are not intended; how to treat new services like callback and calling card services under the modes of delivery; and how

Box 3.1 The General Agreement on Trade in Services (GATS)

Part I Scope and Definition

Article I	Scope and Definition

Part II General Obligations and Disciplines

Article II	Most-Favored-Nation Treatment
Article III	Transparency
Article IIIbis	Disclosure of Confidential Information
Article IV	Increasing Participation of Developing Countries
Article V	Economic Integration
Article Vbis	Labor Markets Integration Agreements
Article VI	Domestic Regulation
Article VII	Recognition
Article VIII	Monopolies and Exclusive Service Suppliers
Article IX	Business Practices
Article X	Emergency Safeguard Measures
Article XI	Payments and Transfers
Article XII	Restrictions to Safeguard the Balance of Payments
Article XIII	Government Procurement
Article XIV	General Exceptions
Article XIVbis	Security Exceptions
Article XV	Subsidies

Part III Specific Commitments

Article XVI	Market Access
Article XVII	National Treatment
Article XVIII	Additional Commitments

Part IV Progressive Liberation

Article XIX	Negotiation of Specific Commitments
Article XX	Schedules of Specific Commitments
Article XXI	Modification of Schedules

Part V Institutional Provisions

Article XXII	Consultation
Article XXIII	Dispute Settlement and Enforcement
Article XXIV	Council for Trade in Services
Article XXV	Technical Cooperation
Article XXVI	Relationship with Other International Organizations

Part VI Final Provisions

Article XXVII	Denial of Benefits
Article XXVIII	Definitions
Article XXIX	Annexes:
	Article II Exemptions
	Movement of Natural Persons Supplying Services Under the Agreement
	Financial Services
	Telecommunications
	Air Transport Services
	Negotiations on Basic Telecommunications
	Negotiations on Maritime Transport Services

Table 3.1 Model schedule of commitments on basic telecommunications

Sector or subsector	Limitations on market access (types of measures to be listed)	National treatment limitations (types of measures to be listed)	Additional commitments (types of measures relevant to possible undertakings)
2.c. Telecommunication services (a) through (g) and (o) Local/long distance/international service: • wire-based • radio-based • on a resale basis • facilities-based • for public use • for nonpublic use	(1) Quantitative limitations/need test applied to the number of service suppliers (including monopolies, duopolies, etc.), total value of transactions, total number of operations, or quantity of output. (2) Example not given. (3) Quantitative limitation on the number of available frequencies to be allowed to foreign service suppliers. Restrictions or requirements regarding the type of legal entity permitted to supply the services (also, a requirement of certain forms of commercial presence could rule out cross-border supply). Limits on foreign equity participation. (4) Limitations/needs test applied to the total number of natural persons that may be employed.	(1) Preferences given to domestic suppliers or restrictions imposed on foreign suppliers in the allotment of frequencies (3) Preferences given to domestic suppliers or restrictions imposed on foreign suppliers in the allotment of frequencies. Limitations on the nationality or residency of directors or board members. Restrictions on foreign ownership of facilities. (4) Example not given.	(Commitments on measures not subject to scheduling under Articles XVI and XVII, including but not limited to those regarding qualifications, standards, or licensing requirements or licensing procedures and other domestic regulations that are otherwise consistent with Article VI and the Annex on telecommunications.) Separation of regulatory and operational functions. Safeguards against anticompetitive practices (i.e., of monopolies and dominant providers). Procedures or requirements related to: ▪ licensing ▪ allotment of radio frequencies ▪ numbering and identification codes ▪ type approval ▪ interconnection Pricing related measures (e.g., cost-oriented pricing: participation in the standards-setting process, including review and comment prior to adoption of new standards. Rights of way for the construction of infrastructure.

Note: Numbers in parentheses indicate the mode of delivery: (1) cross-border supply; (2) consumption abroad; (3) commercial presence; and (4) presence of natural persons.

Source: World Trade Organization.

to deal with video services, particularly with respect to their provision via direct-to-the-home satellite systems.

The biggest conceptual and policy breakthrough in the process was the agreement to create a Reference Paper on regulatory reform as a framework that governments could endorse in the Additional Commitments portions of their schedules. The final paper laid out six key principles for the redesign of national regulatory rules and institutions to ensure compatibility with trade disciplines. These are:

■ *Competitive safeguards.* Governments are required to ensure that major suppliers, especially the national PTOs, do not engage in anticompetitive cross-subsidization, use information gathered from competitors with trade-restricting results, or fail to make available, on a timely basis, the technical information about their facilities and operations needed by competitors to enter the market.

■ *Interconnection.* PTOs are to provide market entrants with interconnection at any technically feasible point in the network. Interconnection is to be provided at nondiscriminatory terms, conditions, and rates, and should be of a quality no less favorable than the provider gives its own services. Moreover, interconnection rates are to be cost-oriented, transparent, and, where economically feasible, unbundled. A dispute mechanism administered by an independent body is called for to handle disputes over interconnection terms and other issues.

■ *Universal service.* Such obligations are to be administered in a transparent, nondiscriminatory, and competitively neutral manner that is not more burdensome than required to meet the policy objectives.

■ *Public availability of licensing criteria.* Where licenses are needed, information and decision-making procedures are to be transparent.

■ *Independent regulators.* Regulatory bodies are to be separated from service providers and not accountable to them.

■ *Allocation and use of scarce resources.* Procedures for allocating and using frequencies, numbers, and rights-of-way are to be carried out in an objective, timely, transparent, and nondiscriminatory manner.

As the NGBT progressed, a number of countries shifted their status from observer to full participant, and offers were enriched in the course of bilateral consultations. By April 1996, 34 offers were tabled by 48 governments (including the European Union), many of them endorsing the Reference Paper. Nevertheless, key US companies were not satisfied that key countries had made sufficient concessions on satellite services. In addition, some facilities-based carriers in the United States (especially AT&T) balked at the potential for asymmetries between countries in the conditions of entry. For example, they argued that carriers from countries with relatively closed markets might get access to the open US market via low-cost resale, while American providers would be left to use the higher-priced accounting rate system for outbound traffic, thereby resulting in one-way bypass. Hence, the United States walked out and the negotiations could not conclude an agreement. However, encouraged by the WTO secretariat, the parties decided to extend the process, with somewhat different procedures.

Hence, a new Group on Basic Telecommunications (GBT) was launched. Between July 1996 and February 1997, nine meetings were held to clarify the outstanding issues and build consensus, again accompanied by informal rump group sessions and intensive bilateral negotiations over the schedules. New countries joined the process, and the offers gained weight. To overcome last minute obstacles, high-level interventions by President Clinton and other national leaders helped smooth over some specific objections by such countries as Argentina, and on 15 February 1997, the GBT was successfully concluded. The Fourth Protocol to the GATS, which provides the legal basis for the codification of the deal, enters into force on 1 January 1998.

Potential Significance of the GBT Deal

Will the GBT deal make a major difference in the global telecommunications industry? Certainly, participating governments' commitments contain various types of exceptions, many of the developing countries' concessions are to be phased in over the next decade or so, and some important countries were not parties to the negotiating process. Yet to focus on these limitations is to see the trees but not the forest. Even though the GBT deal does not require immediate, total liberalization of all market segments everywhere in the world (hardly a realistic prospect, and arguably not a desirable one either), 69 countries accounting for more than 90 percent of the global market nevertheless undertook important liberalization obligations under a coherent multilateral framework. In many (but admittedly not all) cases these commitments represent significant departures that should expand competition beyond previous liberalization programs. Failure to implement the commitments could result in contentious consultations with trade partners and, if these fail, binding dispute resolution. Moreover, they cannot easily be withdrawn or modified, and there will be strong pressures to extend their scope in future rounds of trade talks—progressive liberalization in WTO-speak.

But while they are obviously of great interest to potential entrants, these short- to medium-term changes in market access conditions in various countries are really only half the story. For in a broader sense, the GBT deal represents a significant deepening of the institutional reorganization of telecommunications governance. At the national and multilateral levels alike, governments will have to redefine their policy apparatuses and conceptual frameworks in ways that will matter in many more instances beyond the immediate market-access concessions. Moreover, in government and in the private sector as well, free trade coalitions will be strengthened and may become permanent features of the policymaking landscape. With these factors in mind, the GBT deal should be seen more as the start of a film than a one-off snapshot. And if the results of the negotiations cannot be adequately assessed in static terms, neither should they be viewed in isolation. They will take on even greater weight as they interact with other changes underway in the technological, economic, and political environments in ways we may not be able to foresee today. Hence, if we focus not only on who opened up what markets how much in this particular negotiation but also on the larger implications for how telecommunications policy will be conducted in the future, it becomes clear that the GBT deal could indeed represent a sort of big bang, albeit a slowly building one. What follows are some arguments about why this is so.

Wide Range of Services Covered

During the Uruguay Round negotiations, the definition and classification of the telecommunications services to be made subject to GATS disciplines engendered complex discussions and some disagreements. In the end, two interrelated approaches to these issues were adopted. First, for the purpose of defining the telecommunications services to which suppliers must have rights of reasonable access and use in delivering other scheduled services, the Annex on Telecommunications specifies that "'Public telecommunications transport service' means any telecommunications transport service required, explicitly or in effect, by a Member to be offered to the public generally. Such services may include, inter alia, telegraph, telephone, telex, and data transmission typically involving the real-time transmission of customer-supplied information between two or more points without any end-to-end change in the form or content of the customer's information" (WTO 1994, 360). This broadly framed language is, in effect, a conventional (if perhaps technologically dated) definition of basic telecommunications.

Second, for the purpose of scheduling national commitments, negotiators agreed to use the following classification of telecommunications services:

a. Voice telephone services

b. Packet-switched data transmission services

c. Circuit-switched data transmission services

d. Telex services

e. Telegraph services

f. Facsimile services

g. Private leased circuit services

h. Electronic mail

i. Voice mail

j. On-line information and database retrieval

k. Electronic data interchange

l. Enhanced facsimile services (including store and forward, store and retrieve)

m. Code and protocol conversion

n. On-line information and/or data processing (including transaction processing)

o. Other (e.g., trunked radio, mobile cellular telephone, paging, fixed and mobile satellite services, teleconferencing)

Categories a through g were agreed to be examples of basic services, and only 19 of the 48 national schedules submitted included some highly selective commitments in them. In contrast, categories h through n were deemed to be value-added services, and most of the schedules included multiple commitments in them. Category o was sort of a grab bag, with most of the services given as examples thought to be basic and hence not the focus of extensive commitments. However, there was some variation in how governments classified such services at the domestic level and listed commitments in their schedules, which complicated the process at times.

In the post-Uruguay Round negotiations, governments wanted to sidestep such discrepancies and also to avoid locking in a detailed, singular, and potentially exclusionary definition of basic telecommunications services. Moreover, they sought to establish a technologically neutral deal that did not equate a priori particular services only with specific network delivery options (although of course governments could choose to delimit their offers by, for example, listing global mobile personal communications services or digital cellular services rather than mobile services generally). In a dynamic environment where rapid innovation quickly generates new systems and services that were not envisaged just a short time ago (e.g., Low Earth Orbital Satellites and Internet access) a technology-dependent definition could have unduly constraining effects.

Hence, while a few governments opted to include additional value-added service commitments in their schedules, negotiators agreed to focus on services that—as per the Annex on Telecommunications definition—move customer information between points without value-adding changes in form or content, but to include both services to the general public and private services, for example, those supplied to closed user groups. Commitments were scheduled using the categories employed in the Uruguay Round for each of the four modes of supply. Both networked service provision and resale were on the table, which opened the door to market access commitments for both the commercial presence and cross-border delivery modes of supply. In effect, this acknowledges the reality that trade and investment in telecommunications are closely related and must both be subject to multilateral rules for competition to flourish.

However, while the scope of services negotiated on was broad, and nothing was formally excluded at the outset, some parties drew lines in the sand as the process unfolded. For example, at the behest of France and Belgium, the European Union argued that it would be inappropriate to include mass media transmission services under the rubric of basic telecommunications and therefore refused to schedule commitments in this area. Its schedule stipulates that "Telecommunications services are the transport of electromagnetic signals—sound, data, image and any combination thereof, excluding broadcasting [defined in a footnote as 'the uninterrupted chain of transmission required for the distribution of television and radio program signals to the general public, but does not cover contribution links between operators']. Therefore, commitments in this offer do not cover the economic activity consisting of content provision which require telecommunications services for its transport" (WTO 1997a).[7] Further, due to a dispute with Canada, the United States ultimately decided to take MFN exemptions on one-way direct-to-the-home and direct broadcasting satellite services, as well as on digital audio services. Eight other countries also scheduled horizontal or service-specific MFN exceptions. These exceptions aside, the net result of the process is that a wide range of services in 69 countries are for the first time subject to multilateral trade disciplines.

Significant Market Access Commitments

While the initial offers floated early in the negotiations tended to reflect existing domestic regimes, over time they were enriched to cut deeper into domestic limitations. In this sense, there is a qualitative difference between the schedules produced during the Uruguay Round and in the

---

7. There was even pressure to apply broadcast regulations and limitations on video transmitted over the Internet, which is not a broadcast medium. On the politics of what to include or not in the negotiations, see Beltz (1997, 63-70).

NGBT phase and the additions made to them by the GBT. Previously, governments generally offered standstill concessions; the liberalization programs already undertaken at the national level were simply translated into WTO commitments. During the GBT talks, in contrast, many governments ended up offering rollback concessions that commit them to go further than their existing domestic liberalization programs, albeit for developing countries often on a phased-in basis. This is somewhat ironic in that competition in basic telecommunications has always been seen as more sensitive and politically difficult than it is in value-added services. The shift reflects how far governments have come in their thinking over the past year.

What concessions were made? According to informal tabulations generated by the WTO staff, for voice telephony, 47 of the schedules (covering 61 governments) committed to competitive supply by two or more providers. Generally these allow the supply of public voice services, either immediately or on a phased-in basis, in at least one market segment. Breaking this down further, 41 schedules (55 governments) made commitments on local service, 38 schedules (52 governments) made commitments on domestic long-distance, and 42 schedules (56 governments) made commitments on international service. Resale of public voice services is included in 28 schedules (42 governments). It should be noted that 25 of the 61 governments that committed to some form of competitive supply for voice telephony will phase in their commitments.

For nonvoice services, 49 schedules (covering 63 governments) included commitments on data transmission; 41 schedules (55 governments) allow competition in the supply of leased circuit capacity; 46 schedules (60 governments) allow market access for cellular mobile services; 45 schedules (59 governments) include commitments on other types of mobile services like personal communications services, mobile data, or paging; 37 schedules (51 governments) committed on some or all kinds of mobile satellite services or transport capacity; 36 schedules (50 governments) made commitments on fixed satellite services or transport capacity; and 8 governments scheduled additional commitments on value-added services like electronic mail, on-line data processing, and database retrieval (WTO 1997b, 1).

Slicing the numbers from another angle, the US Trade Representative's office concludes that 29 governments—two-thirds of them from industrialized countries—guaranteed market access for international telecommunications services and facilities. Another 23 governments, primarily from the developing and newly emerging countries—will phase in such commitments, mostly by the year 2006. Six additional countries are open for only selected international services, while 12 have little or no market access commitments in this area. Regarding commercial presence via direct foreign investment, 27 countries (including almost all of the industrialized world) permit foreign ownership or control of all telecommunications services and facilities. Exceptions here include Australia (Voda-fone and Telestra), Belize (the state-owned company), Chile (local service), France (France Telecom), Italy (Stet), Japan (KDD and NTT), New Zealand (49.9 percent limit in Telecom NZ for any one foreign firm), and Spain (Telefonica). Another 21 countries, mostly in the developing world, phase in investment commitments by the year 2004. Ten countries permit more limited foreign ownership or control of certain networks and services, while 12 countries allow only minority equity stakes (USTR 1997).

While it is beyond our purpose here to review all of the 55 schedules in detail, some broad patterns should be mentioned. Most of the industrialized countries have committed to provide market access via the relevant modes of supply for all basic services and market segments (local, long-distance, and international). The most common exceptions involve limits on equity ownership (especially in traditional national carriers) that have been retained by some European countries, as well as by Canada, Japan, and South Korea; and phased-in liberalization schedules for less affluent countries like Spain, Portugal, Greece, and South Korea. In many cases, these countries were already well on their way toward opening markets, although a number of the commitments exceed previously announced initiatives and lock them into international law. In other cases, most notably South Korea, the agreement took the government far beyond what was previously announced.

Arguably, the most notable commitments came from other regions. With regard to the so-called newly industrializing countries, Chile and Mexico—which, admittedly, were already liberalizing—went the furthest by committing to full competition, save local telephony in Chile and some equity limits in Mexico. Others, like Brazil, India, Indonesia, Malaysia, South Africa, Thailand, and Turkey made more carefully circumscribed (and sometimes very limited) offers that opened certain market segments while reserving some or all of voice telephony for companies that had already been granted exclusive rights. In some such cases, promises were often made to review the situation after the year 2000 or the passage of new national laws.

The emerging markets of Eastern Europe—Bulgaria, Hungary, Poland, the Czech and Slovak Republics, and Romania—also undertook notable but limited commitments and promised to open themselves to further (usually complete) liberalization between 2000 and 2004. And some of the biggest surprises came from the lower-income developing countries. El Salvador clearly went further than the rest by offering full competition immediately, an initiative well exceeding prior liberalization efforts. Most of the others offered the usual liberalization of specific markets with full or nearly full competition to be phased in later, for example, Bolivia in 2001, Cote d'Ivoire in 2007, Grenada in 2006, Jamaica in 2013, Mauritius in 2004, Morocco in 2001, Peru in 1999, Trinidad and Tobago in 2010, and Venezuela in 2000.

Of course, one could argue that the various limitations and phase-in periods mentioned above mean that the agreement does not immediately result in a harmonized and rigorous level of liberalization across all participating countries. Inevitably, GBT members held back in certain instances for a variety of reasons—to preserve regulatory flexibility for new circumstances, to give their domestic industry time to adjust before facing new competition, to cater to local interest group pressures and conceptions of the public interest, and so on. Even so, the commitments that have been made remain notable in a great many instances, and pressures for progressive liberalization will inexorably spill over from the open market niches to the less open ones in the years to come.

Widespread Adoption of Proliberalization Regulatory Principles

According to the WTO staff calculations, a surprising 63 of the 69 governments submitting schedules committed themselves to new domestic regulatory principles that are subject to international monitoring, consultation, and dispute resolution. A full 57 countries committed themselves in whole or with few exceptions to the Reference Paper's language on competitive safeguards, interconnection, universal service, public availability of licensing criteria, independent regulators, and allocation and use of scarce resources. The number of countries agreeing to endorse the Reference Paper and appending it to their schedules was one of the most dramatic leaps forward achieved during the GBT negotiations. In April 1996, when the NGBT ended unsuccessfully, only 44 governments had included regulatory commitments, and only 31 had signed on to the Reference Paper.

The incorporation of regulatory principles into a trade policy framework was a remarkable achievement. After all, the entire history of global telecommunications has involved states regulating their national systems as each saw fit. The instruments of the international regime negotiated in the International Telecommunication Union (ITU) positioned sovereignty as an overarching principle, and governments sought to insulate their markets from foreign influence while reaping the benefits of international correspondence, for example, by interconnecting monopoly systems at designated gateways and jointly providing services on a noncompetitive basis.

With the shift to a trade framework of global governance, the new forms of international service delivery require what have been called beyond-the-border and deep-integration measures. GBT members have bound themselves to make their domestic regulatory institutions and rules consistent with multilateral trade disciplines and transnational market forces. It is easy to imagine that much of the agreement's thrust might have been frustrated without these regulatory obligations. For example, offering market access via commercial presence or cross-border delivery might mean little without an internationally recognized right to interconnect with other public telecommunications networks, or without transparent licensing criteria.

Implementation of the Reference Paper's principles on competitive safeguards, interconnection, transparency of licensing criteria, independent regulators separated from the service provider, and allocation and use of scarce resources will be a very demanding task, especially for the developing countries. The industrialized countries and multilateral agencies like the World Bank and the ITU will have to work closely with them to provide the requisite technical assistance. But if new regulatory frameworks are successfully institutionalized, this will affect how all policy decisions and industry plans will evolve in participating countries well into the future. In this sense, the implications of the agreement extend far beyond the market-access concessions that were made in the GBT.

Application of GATS General Obligations to Basic Telecommunications

The deal is significant not only because of the breadth and depth of the liberalization commitments listed in the national schedules, but also because the GODs of the GATS Framework Agreement apply to the national commitments. As such, much of the telecommunications industry has been brought fully under the trade mechanisms of the WTO; indeed, telecommunications has become one of the best-covered service sectors in the GATS.

All the GODs are important, but five merit particular mention here. Article II on MFN treatment entails a big shift away from the reciprocity policies that frequently generate friction in international trade relations, in particular between the United States and its trade partners. For companies seeking entry into markets like the United States, nondiscrimination (save where specific exemptions are scheduled) could represent a substantial change from the status quo ante. And indeed, some movement away from strict reciprocity may soon be visible. In June 1997, the Federal Communications Commission (FCC) tentatively concluded that it should eliminate its controversial "effective competitive opportunities" test as part of the public interest analysis undertaken when carriers from WTO member countries seek to provide facilities-based, resold switched, and resold noninterconnected private line services (FCC 1997, 97-195).

Article III on transparency requires the publication of measures of general application and prompt replies to requests for information, and sets notification requirements concerning new measures and changes to existing measures. These requirements will involve significant adjustments in the anticompetitive practices of many telecommunications ministries and carriers. Article VI on domestic regulation may be equally significant. It requires governments to maintain or establish tribunals or procedures for the prompt and impartial review of complaints; to ensure that all measures are administered in an impartial manner; to advise service suppliers within a reasonable time of their decisions as to permitted operations; and to refrain from instituting new practices that are inconsistent with GATS principles.

Article VIII on monopolies and exclusive service suppliers binds governments to ensure that PTOs do not, in the supply of reserved services, act in a manner inconsistent with MFN and their Specific Commitments. The same is true for abuse of dominant positions, for example, cross-subsidization of competitive services from reserved services, and for the negotiation of compensatory adjustments when monopoly rights are newly granted. Finally, Article IX states that in the case of business practices that restrain competition and trade, governments must, upon request, enter into consultations with a view to eliminating such practices.

The application of these GODs to basic telecommunications could signal a major shift in the dynamics of market entry and scope of competition. For example, a government cannot, with impunity, schedule the removal of quantitative limitations on commercial presence in packet switching but then layer on a set of procedural hurdles and information restrictions that make access all but impossible. This sort of openness in name only has often characterized reform initiatives, and it has allowed the accumulation of significant cross-national asymmetries in the actual (as opposed to announced) level of liberalization. Practices that derogate from the applicable GODs can now be challenged in bilateral consultations, and cases that cannot be resolved may move to the WTO dispute resolution mechanism.

Institutionalization of Multilateral Surveillance and a Framework for Consultations

This last observation leads to a broader conclusion that frames the points made below. From an institutional perspective, the real significance of the GBT deal does not rest on how deeply countries have liberalized any given sector or subsector in the short term. What may matter more for the governance of the global information economy is that the deal signals the beginning of an evolutionary process of mutual adjustment that will unfold according to a clearly defined set of principles, baselines, and mechanisms. For the GBT agreement institutionalizes a multilateral system of mutual surveillance and a framework for bilateral consultations on implementation issues.

Regarding the former, the deal extends GATS mechanisms to provide a variety of information-sharing procedures through which everyone can see what everyone else is doing in basic telecommunications and judge whether their actions conform with shared principles. Through such mechanisms as meetings of the telecommunications group of the Council for Trade in Services, reports on country conditions, and bilateral interactions, the ability of member governments to hide violations of their international commitments should be curtailed. Regarding the latter, clear, substantive rules are established according to which bilateral consultations can be conducted. A country whose service supplier is seeking entry abroad is no longer simply left to plead with its trade partner. Instead, it can now claim a right based on a binding and explicit set of shared guidelines.

The net effect of these surveillance and consultation mechanisms should serve to create a measure of convergence in the pace and content of national liberalization programs. That could be significant for the difficult implementation work that lies ahead.

Institutionalization of Dialogue and Conceptual Convergence

The GBT agreement also institutionalizes dialogue among parties and collective examinations of problems, thereby promoting conceptual progress and convergence. All social actors—individuals, businesses, governments, etc.—behave on the basis of cognitive constructs, beliefs, interpretations, and understandings of their interests and the larger environments in which they operate, and these mental images help to define what is appropriate in any given situation. As many social scientists have shown, the development of cognition generally is not something that happens in an atomistic, entirely individualized manner, since most of us do not live in isolated caves. Instead, it is to a significant degree a collective experience, one that plays out through complex processes of ongoing interaction and communication. The collective development and internalization of shared information and beliefs is an essential underpinning of all international institutions and patterns of world order. Indeed, the institutionalization of trade-in-services discourse in the 1980s signaled a major shift in governments' collective understanding of international telecommunications and in the norms for defining, bargaining over, and resolving policy problems.

The GBT deal expands this process of conceptual development and convergence. The very act of having to think through their market access and national treatment policies for each service on each of the four modes of supply was an important exercise for governments to go through. So too was the elaboration of the Reference Paper's principles, which each government had to think about in relation to its domestic regulatory regime and market structures. The utility of these processes was especially great for developing countries, which previously often lacked clear policies on all the questions involved. They will need a great deal of assistance in redefining their domestic regulatory systems to achieve conformity with international standards and notions of best practice. Dialogue on these and related matters will now continue in an ongoing, structured manner in the years ahead. Further, at a more mundane level, questions of statistical measurement and national accounting of basic services also will be continuously addressed in a manner that promotes shared understandings and negotiations.

Dispute Resolution and Sanctions

The GBT deal brings basic telecommunications—and hence, most of the global market—under an internationally accepted enforcement mechanism. For countries that wish to fend off bilateral pressure tactics, this represents an important break with the past. The institutional mechanisms may be especially important for small and developing countries that would otherwise find themselves at a disadvantage in dealing with more-powerful countries. The process begins with bilateral consultations in which the Director General of the WTO may offer his good offices to mediate the dispute. If these efforts fail, the Dispute Settlement Body (DSB) will establish a panel with clear terms of reference and an agreed composition. The panel then examines the issues in meetings with the parties and concerned third parties over a period not to exceed six months (three months in urgent cases). The panel submits its report to the parties and the DSB for an interim review. The DSB must decide on the report within 60 days. An appellate review not to exceed 90 days may then be launched. The DSB monitors the implementation of the adopted panel report or appellate body recommendation, and parties may then negotiate compensation pending full implementation. If all else fails, the DSB can authorize retaliatory measures against the infringing party. Never before have telecommunications services been subject to such a clear and forceful multilateral framework for preventing and resolving bilateral conflicts; the instruments of the ITU, for example, lack provisions for settling disputes over competition. Trade policy officials hope that the carefully staged sequence of WTO dispute resolution steps will resolve differences before they reach the boiling point.

There are, however, potential problems with this framework. One is whether the composition and conduct of the DSB and its panels will prove sufficiently open and transparent to command widespread respect. Another is whether the involvement of the DSB and the WTO telecommunications group in the intricacies of domestic regulatory issues like interconnection will generate new conflicts. Finally, there is no guarantee that the sequence of dispute remedies, up to and including retaliation, will actually change offending practices when powerful countries find themselves on the losing side of a DSB decision.

Consolidating Proliberalization and Free Trade Coalitions in Member Countries

The GBT deal helps to consolidate proliberalization coalitions. The institutionalization of trade in services as a subject of negotiation over the past decade has been an important factor in altering national policy-making processes. In the past, telecommunications policy was largely the proprietary preserve of ministries of posts and telecommunications (or independent regulatory agencies like the Federal Communications Commission), national carriers, and other powerful interest groups that typically had stakes in the status quo. When national carriers were closely linked to or part of the ministries, government policy not unsurprisingly acted to preserve their monopolies or at least slow down competitive entry. Similarly, even independent agencies like the FCC frequently succumbed to strong influence from the companies they were to regulate; some observers go as far as to allege that the FCC has often been subject to capture by dominant firms in telecommunications and broadcasting, although the historical evidence for this is less than consistent.

Either way, absorbing telecommunications into the trade policy environment has clearly altered the mix by making the organizational objectives and intellectual frameworks of trade and finance ministries a vital part of decision making. Permanent bureaucratic coalitions have been established with stakes in promoting liberalization to the benefit of the economy as a whole, rather than in protecting the prerogatives of traditional national carriers. In Europe, this has also strengthened the position of the European Commission relative to national telecommunications ministries. Further, the new coalitions help raise awareness of telecommunications policy among ruling politicians, frequently driving the issues closer to the top of the national agenda.

The same sort of thing has occurred in the private sector. Proliberalization firms that are looking to international markets have been emboldened to press for more openness at home as a price for gaining access abroad. Users, too, have pushed for freer trade, since this allows them to purchase the best service without regard to the nationality of the provider. Telecommunications ministries, PTOs, protected manufacturers, and other players whose first preference might be for closed markets can no longer unilaterally define problems and get their way without being challenged.

The GBT deal deepens these trends. It does this most directly by requiring, in line with the Reference Paper and the GODs, that carriers and ministries be separated, and that transparent and fair procedures be employed in dealing with foreign requests for market access, authorization, and so on. But it does this indirectly as well, by giving trade and finance ministries and protrade firms a larger voice in the policy mix. And this is true not just in matters of value-added services and private

networks, as before, but now also in the biggest portion of the market—
basic telecommunications.

Redefining Multilateral Governance

The domestic shift in the balance of influence has a multilateral counter-
part. Since the late 1980s—and especially since the 1988 World Administrative Telegraph and Telephone Conference—the ITU has changed its spots.
Gone are the days when the Place des Nations complex served as the
exclusive clubhouse of national monopolies that designed international
agreements to buttress their market power and preclude competition. The
organization has become infinitely more open and representative of the
growing diversity of global telecommunications stakeholders. The ITU's
institutional procedures have been adjusted to actively solicit the participation of a wide variety of private firms in its work and decision making.
   Moreover, many of the key instruments of the international telecommunications regime—e.g., the accounting and settlements system, the
technical standardization process and its products, the regulations of private leased circuits and networks, and so on—have been adapted to
allow governments and firms to develop competitive market arrangements. Indeed, the regime has undergone a fundamental transformation
over the past decade.[8]
   The entry of the GATT, now the WTO, into the global policy mix—
which has been called the ultimate bypass—has been an important part
of this process over the past decade. The shift to a trade framework has
meant that some of the most interesting and important activity has moved
down the street from the ITU's Place des Nations site to the WTO's
location at Place Albert-Thomas. Together with the growing involvement
of other regional, plurilateral, and multilateral institutions, the WTO's
growing role means that jurisdiction over key aspects of global governance has become a contested market. The ITU continues to play
essential roles on many issues—e.g., technical standardization, frequency
spectrum management, assistance to developing countries, and so on—
but the locus of debate over the main economic issues has been moving
to the WTO. This is significant because none of the ITU's instruments
actively promotes competition and market entry; they simply allow it, if
members wish to follow that route.
   The GBT deal deepens these trends. As the Council for Trade in Services' work and the consultations on implementation issues take off,
governments and firms in the ITU at times will have to adjust their
programs and policies to accommodate the WTO process. Technical stan-

---

8. For discussions of these developments, see Drake (1994a, 137-203); Drake (1994b, 71-
96); Rutkowski (1995, 223-250).

dards, radio frequency spectrum management agreements, and other instruments negotiated in the ITU may need to be examined in relation to WTO commitments, and there may be difficulties in certain instances. No wonder that in June 1997, the ITU's Council decided to schedule a second World Telecommunication Policy Forum for March 1998 to consider the implications of the GBT deal and related issues.

One particularly important effect of the GBT deal may be on the accounting and settlements system that has been developed in the ITU and widely implemented by national carriers. Efforts within the ITU over the past six years to reform the system and move toward cost-oriented pricing, nondiscrimination, and transparency have been very slow to yield results. By fostering the development of competition on international routes, including corporate alliances and new forms of direct interconnection, the GBT deal should encourage a growing share of international traffic to move outside the accounting rate system. This could provide fresh incentives to reduce accounting rates, and alternative arrangements like termination fees and facilities-based interconnection charges may become more potent competitors. In short, together with new technologies and services like callback and internet telephony, the deal may well be a factor in undermining this market distorting part of the international telecommunications regime and fostering a broader restructuring of the global communications order.

Potential Problems With the GBT Deal[9]

Politics is the art of the possible. By that standard, the world trade negotiators deserve a pat on the back for passing the GBT deal, after 10 years of trying. The agreement is a step in the right direction. But it is quite another matter to declare it, as credit-grabbing victory bulletins did, a revolution, a breakthrough, a telecommunications D-Day. Much of that view is steeped in the belief that reality in the information sector is shaped by intergovernmental rules, rather than the other way around.

Of course, the people directly involved in the drafting, lobbying, analyzing, and implementing of the agreement have worked hard to seal the deal, and it is therefore natural for them to believe that the result of their attention has been a monumental change rather than a monumental effort. And because the trade community is influential, this negotiation-centric view of the world becomes the common perspective. Doubts are dismissed as protectionist, as if government agreements necessarily further free markets. Or, it is argued, doubters seek perfection in a messy world, whereas the agreement merely initiates a process. But

9. The assistance on this section of Marc Austin, John Kollar, Jennifer Schneider, and Gene Fang is gratefully acknowledged.

that process has been going on for years, and would have continued. Thus, the question must be asked: Does the WTO's basic telecommunications deal make much of a difference? Is it a cause or merely an effect of more fundamental change? And the answer is: The impact of the agreement is certainly much less than claimed. In some areas, it will make no difference to the process of liberalization. In others, the deal modestly accelerates processes that had started already. And in still other cases, it may even slow down future change. This will now be discussed.

The Scope of the Agreement Is Being Exaggerated

The offers of market access by the various countries, according to the WTO, account for over 90 percent of telecom revenue worldwide. Some estimates claim up to $1 trillion in gains to global income. According to the US government, the agreement will lead to an 80 percent reduction in the costs of telephone calls and create up to a million new jobs in the United States alone. Similar pronouncements were issued by other countries and repeated by an underinformed press. But are they correct?

Some skepticism is in order. Let us begin with the scope of the agreement. It is less remarkable than numbers suggest. A total of 73.6 percent of the world telecommunications market is accounted for by the United States, Japan, and the European Union. These countries had already committed themselves to liberalization and market opening as a result of their own internal evolutions. They were also in the process of opening to each other. To mention but two of numerous examples, BT had already announced its purchase of MCI, and Ameritech was part owner of Belgacom. Many other countries have been similarly moving toward liberalization and market opening on their own, including Australia, Mexico, New Zealand, Canada, and Hong Kong.

Furthermore, simply adding the tele-populations is also misleading, because the GBT agreement is not a treaty in which the signatories agree to the same set of actions. It is more in the nature of a series of pledges. The scopes of national commitments differ greatly. Many countries attached conditions to or delayed the date of their implementation, thereby in effect providing a protective or protectionist cushion.

Foreign investment commitments are an example. A total of 56 countries made such commitments. Of these, 21 countries are committed to implement the agreement at a later date (from 1999 for Peru to 2004 for Pakistan), and 18 have limitations on foreign investments (either on incumbents or on selected services). For example, Canada maintained its current limit on foreign carrier ownership of 46.7 percent, including a limit of 20 percent for direct ownership in facilities-based suppliers. Japan maintained 20 percent ownership limits for NTT and KDD. The major European countries reserved the right to maintain state holdings in their national carriers, including France Telecom, Deutsche Telekom, and Telefonica of Spain. In Portugal, only companies established in the country will be able to offer international services. Only 17 nations have made foreign investment commitments without reservations.

Of the offers covering voice telephony, 40 percent are to be phased in over various periods of time (WTO 1997a). Most of the early adoptees are the major telecommunications countries already engaged in such liberalization.

Among excluded services, the transmission of audiovisual material is particularly important. As telecommunications and broadcasting converge through digital technologies, the audiovisual exclusions could become a huge loophole in liberalization. Internet-style video, for example, would be excluded by countries that define it as a one-way service to the public.

The deal's exemptions and exceptions are not the end of the story. In implementing it, democratic countries will not meekly follow their trade negotiators. The US Congress, for example, has been vocally concerned about foreign investment restrictions in Mexico, Canada, and Japan, and may well intervene. Already, the United States has begun to undermine the WTO procedures in its dispute with Europe over the Helms-Burton law dealing with Cuba. The United States argued that the matter affects its national security and is thus exempt from panel adjudication. Similar exemptions could be claimed, in telecommunications, for a variety of reasons, for example, environmental protection or employment. The WTO panels could ignore such arguments, but this points precisely to their problem in terms of democratic legitimacy—they are likely to be single-issue-oriented experts focused on the trade dimension. The political process is unlikely to accept such abrogation of powers—not to mention sovereignty—and will therefore continuously add conditions affecting countries' implementation.

The treatment of satellite communications shows the hollowness of many of the offers. This segment of telecommunications is potentially the most instantaneously open segment to international entry. It is also the segment where many countries chose to delay market access from the normal 1 January 1998 deadline. Argentina, for example, limited fixed satellite service via geostationary satellites in order to protect its satellite operations. Brazil required firms to use foreign satellites only when they "offer better technical, operational, or commercial conditions" (A Review of WTO Commitments, *Communications Daily*, 1 May 1997). Of the 52 countries that agreed specifically to include satellite services in their offers, 23 chose to delay access. Canada, for example, agreed to eliminate Telsat Canada's exclusive rights on satellite facilities and earth stations serving the North American market only by the year 2002. Partly in response, the United States took an exemption from providing MFN treatment for digital broadcasting, direct-to-home video services via satellite, and digital audio radio services. It argued that these services are defined in the United States as telecommunications, but in other countries as broadcasting (which is outside the deal), which would therefore create asymmetrical discrimination against US providers in other countries.

Thus, the most likely competitive entry route by international carriers—satellite communications—is on a slow track for most countries, whereas other and less likely entry routes are becoming formally open, possibly because such entry poses no competitive threat. It was, of course, always possible for countries to delay opening to satellite competition. But with the deal, such delay has become part of an internationally agreed-upon framework, and there are fewer grounds to complain.

Liberalization and Market Opening Were Happening Anyway

As mentioned, an increasing number of countries had already initiated various unilateral and bilateral liberalization initiatives and would have continued to do so. The United States has been liberalizing at home at an accelerating pace, and has been pushing the same agenda internationally. In Europe, the European Commission has moved its members toward liberalization, privatization, and market opening. Europe had internally agreed on substantial liberalization by 1998. The GBT agreement did not advance that schedule.

The developing world, too, turned away from state monopolies. Much of Latin American telecommunications has been privatized already, motivated by the need to lower debt burdens and by efforts to push firms and bureaucracies toward economic reform. Often, developing countries liberalized to attract foreign investment. Even before the GBT agreement, international opening has generated, in a few short years, a truly astonishing number of transborder telecommunications investments, with no end in sight (Noam and Singhal 1996, 769-87). Several of the largest carriers have already achieved a significant penetration of international markets. According to the FCC, in 1996 BT was present, directly or by way of venture activities, in markets accounting for 80 percent of multinational corporations and 57 percent of all international voice traffic (Galbi and Keating 1996). AT&T has a similar, but lower, market penetration. GlobalOne, another of the main alliances, was said to have an even greater reach. The GBT increases the opportunities for such deals marginally, but further opening would have continued in any event.

This is not to say that there will be no effects at all of the deal. Some reluctant countries may have been pushed to go a bit further than otherwise. But since they might still be able to retract in the implementation phase, the jury is still out.

The WTO May Slow Down Liberalization

In the WTO, countries that pioneer policy reform are likely to find themselves in the minority. The WTO will be dominated by coalitions of countries that play the international game well and take a centrist position on reforms, with an emphasis on stability, gradualism, and international compatibility. This could hold back change by reducing national experimentation and by raising the transaction costs of reform.

The notion of the WTO as an agent of change may be wishful thinking. In telecommunications, the history of international agreements, going back to the nineteenth century, has been one of cartel protection. It would be surprising if the new supranational regulatory arrangements were not similarly captured over time by those desiring stability rather than change. In the negotiations leading up to the WTO agreement, large firms influenced outcomes, while few nongovernment organizations or consumer organizations took an active role. This will continue. The complexity and proliferation of WTO telecom trade disputes will leave only the largest of players able to afford continued participation.

The life cycle of regulatory institutions—from youthful vigor to senescent status quo protection—has often been commented upon. Why should the WTO be different, after its early burst of energy is dissipated?

The ITU, the WTO's sister organization, has been captured by the PTOs since its birth more than a century ago, even though this has improved in recent years. There is no reason to believe that similar forces will not accumulate and affect the WTO. Some argue that the WTO is more user friendly and less specialized than the ITU, and hence less susceptible to stakeholder pressure. But in reality, the WTO is just as specialized, only along a different dimension. Trade is important, but surely it is only one factor in a complex society. Those stakeholders most affected will make the greatest investment in the process of influencing the institution.

In institutionalizing that specialization, the WTO creates a distance to politics. There is no openness of the dispute resolution process, no hearing mechanism, no public record. Considerations other than trade could be disregarded. All this will inevitably lead to backlash.

The WTO Process Itself Will Be Slow

Given the 10 years of deliberations needed in coming to an agreement, there is little reason to believe that the WTO telecommunications decision-making process will be speedy. The WTO, as a multinational bureaucracy, will respond to the disputes associated with rapid change in the telecommunications industry—such as callback and internet telephony, to mention two recent examples—even less quickly than most national regulators. Its processing capacity is limited. Inevitably, yet another layer of telecommunications bureaucracy will be created.

It is claimed that the new process, by setting a variety of time lines for the resolution of disputes, will speed up liberalization. Is that true? There is indeed an agreed-upon time schedule, but it applies only to the WTO segment of a dispute. Even there, administrative bottlenecks may create problems and stopped clocks and multistage decisions. More importantly, there may be lengthy procedures in advance of a case reaching the WTO.

The WTO is an organization of governments, not a civil court. Unlike bringing their grievances to a national court, companies do not present cases, governments do. Companies must rely on their governments to put forward their complaints in an intergovernmental procedure. The firm must convince its national government to back its claim and present it to the other governments, the arbitration panel, etc.

This means that (1) the government will have to conduct its own investigation into the complaint; (2) it will next have to make sure that its support of firm A will not negatively impact the trade position or well being of another national firm B; (3) it may even need to make a ruling on whether the firm is in fact a national firm or a foreign entity; and (4) it will have to decide whether the case is worth spending national resources in a protracted WTO battle.

The US government already has made it clear that it will not take all complaints to the panel, but only especially grievous incidents against a single signatory to the overall GATS agreement, and not necessarily in the telecommunications sector.

Even after a WTO panel decision, a company may not receive satisfaction. In the WTO, negotiating bottlenecks may be resolved by linking issues. Concessions in one sector can be traded off against concessions in another. Thus, even after a panel decision, a government can choose, in theory, to pair off acquiescence to a restriction in telecommunications in another country in return for the acceptance of a restriction of its own over, say, toothpaste.

Contributing to slowness is the inevitable jurisdictional dispute among the WTO, the ITU, and other multinational bodies. This is inevitable, despite good-faith declarations of intended collaboration by the principals, because the various interests will invariably engage in forum shopping to find a congenial result. For example, trade barrier issues (the WTO's mandate) could be redefined as technical standards issues (the ITU's mandate).

This is predictable. Interest groups pragmatically desire the regulatory level and body whose outcome they like best, regardless of their official ideology. In the case of the WTO, various PTOs with international ambitions, and in particular large users of telecommunications services, perceived that they would get better results from the WTO to force a few slow but profitable countries to speed up liberalization, and to extract some concessions for steps they would have taken anyway.

All of this means that the dispute resolution process may require a long and expensive road to justice for those firms with legitimate complaints.

It is claimed that the formalized WTO structure will facilitate the exchange of information on countries' policies. Perhaps. But such exchange already takes place in the numerous international conferences, trade shows, and trade magazines and consulting reports. It does not require the existence of the WTO for France to find out about South Korea's interconnection policy.

The Problems of International Agreements

A fundamental question in evaluating the GBT is whether agreements on regulatory policy across international boundaries are necessarily a good thing. Jagdish Bhagwati, one of the world's most respected trade economists and one of the WTO's own senior advisers, has argued that the effort to harmonize regulations and standards among trading nations is hopeless and counterproductive. Free trade is most efficient when there are differences among nations that can be exploited by industry seeking to specialize. When nations seek to harmonize their regulatory environments, they remove many of the gains from trade that we would have in a less rigidly ordered world (Bhagwati and Hudec 1996).

The stated justifications are to facilitate interaction, eliminate negative externalities, prevent free riding, and gain economies of scale (Leebron in Bhagwati and Hudec 1996). But countries differ in endowments, technologies, preferences, institutions, and coalitions. Why then should they share a regulatory approach for telecommunications? The WTO and its GBT are just another instance of the permanent struggle between centralism and diversity, between globalism and localism, between state control and market forces. Each has some advantages, and it is far from clear that a policy agreement around the world is the better way, even for those seeking liberalization.

Why not differentiate telecommunications policy? After all, telecommunications encroach on domestic politics, touching on sensitive issues such as a nation's control of its communications infrastructure, the funding of social objectives, redistribution and employment policies (Beltz 1997, 64), to name just a few examples. Different countries have different tastes for the trade-offs among these issues.

It is argued that there is nothing to prevent a country from becoming more deregulatory than the WTO framework. True. It is also claimed that the agreement puts pressure on countries not to be out of line in liberalizing. Also true, and in conflict with the first statement. By the same logic of international compatibility, the agreement would strengthen those opposed to far-reaching reforms that are out of line internationally. Hence, a possible effect will be to institute a certain parallelism of change—speeding up the slow countries, slowing down the leading edge.

Much is made of the acceptance of a regulatory reference model, making it seem like the adoption of some universal charter of telecommunications freedom. The reality is more modest. The model principles are mostly procedural, not substantive. They speak of independence of the regulator, but this merely refers to independence from the monopolist, not from politics—as if formal independence prevents capture. The principles speak of openness, public licensing criteria, transparency, and objective allocation procedures. All this sounds good, but is worth little because of its vagueness, if a government drags its feet. For example, an openness of process can mean very little outside of formal meetings, whereas most serious work gets crafted outside the public sunshine on the senior staff level before ceremoniously reaching the official decision event.

Entirely missing from the Reference Paper is a principle against public and private monopolies in telecommunications. All the agreement does is prevent discrimination against one set of rival entrants and in favor of another, or its unfair extension into competitive segments. But there is no per se legal restriction against a natural monopoly.

The most substantive aspects of the model rules are commitments to a nondiscriminatory universal service burden (had anyone disagreed before on that issue?) and interconnection. That is indeed important, but it is entirely derivative of the existence of competitors. And the model principles do not require competition. They merely require, if competition is permitted, that it be nondiscriminatory, including in interconnection, cross-subsidization, and access to information. This is useful for those situations where a country permits competitive entry on paper but somehow wishes for the entrants, especially foreign ones, to fail. This is useful, but not critical. The experience of liberalizing countries has been that once a pro-competition policy was adopted, new entrants actually received some form of regulatory protection against the massive power of the incumbent. The reason is that after embracing a liberalization policy, governments do not want it to fail (unless such liberalization is purely the result of foreign pressure).

There are no shortcuts to the internal dynamics of overcoming the traditional supplier monopoly by entry and strengthening the role of the user community.

Conclusion

As the above discussions indicate, there are distinctly different arguments one can make about the GBT deal's implications for global telecommunications. From a skeptic's standpoint, the benefits of any new international collaboration must be weighed against the transaction costs incurred, and in particular against the risk that multilateralism may reduce policy innovation at the national level.

In many cases, the best coordination mechanism would be through market forces and arbitrage, rather than through intergovernmental collaboration. It is true that market forces by themselves do not deal with all policy problems, such as redistributive goals, negative externalities, law enforcement, and the transition to a competitive system that may require interconnection arrangements. But these are primarily national issues, calling for differentiated national responses. These national policy responses, in turn, would create opportunities of entry or prevent them. And the differentiation in offerings and pricing creates opportunities for arbitrage and other mechanisms to whittle at national restrictions. This process may seem less elegant than a set of multilateral agreements. But the implementation of the latter will not prove to be elegant, either.

Thus, from this skeptical standpoint, the GBT deal needs to be seen in the proper perspective. It will unleash very little that is not already in motion due to the fundamental transformation of global telecommunications. It is the result of these forces, not its trigger. It may formalize the process a bit, but it is not clear whether formality of process will drive change or slow it down. It creates a supranational mechanism, but it is not clear whether supranationalism is good for telecommunications reform. In the past, it was not. The agreement pushes some countries toward some reforms. But the effectiveness of liberalization and the significance of telecommunications to economic development had already become too clear to be resisted, even if unilateral reforms are repackaged as multilateral trade concessions. It provides an international forum more congenial to large users, but those organizations would have found other ways to make their voices heard. On the whole, therefore, the deal, while useful in some aspects and toward some countries, will turn out to be modest in impact.

Alternatively, one could argue that the GBT does, in fact, have the potential to be extremely significant for global telecommunications. In this view, there is no reason to believe that the WTO's role will prevent governments from pursuing innovations in national policy, at least not innovations that are consistent with free trade and liberalization. More generally, if one takes immediate, total liberalization of all market segments everywhere in the world as the baseline against which to measure the degree of change, then of course there are grounds for suggesting that the commitments do not go far enough, multilateral cooperation is not fast enough, and so forth. But this is not an appropriate starting point of analysis.

For more than a century, telecommunications has been subject to extensive state control at the national level, while at the multilateral level there have been no rules obliging governments to allow foreign companies to compete. Indeed, the collective institutional frameworks of the traditional global communicatiodns order—the international regimes for telecommunications, radio spectrum management, and satellite systems—were designed to limit or preclude foreign competition. Now, that old order is breaking down (a process that has been gathering momentum for some time) and the GBT deal adds speed to its fall. It applies market opening trade principles to basic telecommunications services in 69 countries accounting for 90 percent of the global market; it will spur governments to undertake deep, proliberalization regulatory reforms; and it institutionalizes multilateral mechanisms that allow governments to pursue alternative policy approaches, but within an agreed set of parameters and under conditions of transparency.

There is widespread hope that the GBT deal will help to establish a new architecture for global telecommunications policy that facilitates increasingly deep liberalization in the years to come. Whether that hope becomes a reality will depend on how it is implemented in national policy, multilateral agreements, and market decisions over the years to come.

References

Barshefsky, Charlene, and Reed Hundt. 1997. Remarks prepared for the House Subcommittee on Telecommunications, Trade, and Consumer Protection. http://www.house.gov/commerce/telecom/hearings/031997/witness.htm (19 March).

Beltz, Cynthia. 1997. Global Telecommunications Rules: The Race with Technology. *Issues in Science and Technology* (Spring).

Bhagwati, Jagdish, and Robert Hudec. 1996. *Fair Trade and Harmonization: Prerequisites for Free Trade?* Cambridge, MA: MIT Press.

Drake, William J. 1993. Territoriality and Intangibility: Transborder Data Flows and National Sovereignty. In *Beyond National Sovereignty: International Communications in the 1990s*, ed. by Kaarle Nordenstreng and Herbert I. Schiller. Norwood: Ablex.

Drake, William J. 1994a. Asymmetric Deregulation and the Transformation of the International Telecommunications Regime. In *Asymmetric Deregulation: The Dynamics of Telecommunications Policies in Europe and the United States*, ed. by Eli M. Noam and Gerard Pogorel. Norwood: Ablex.

Drake, William J. 1994b. The Transformation of International Telecommunications Standardization: European and Global Dimensions. In *Telecommunications in Transition: Policies, Services, and Technologies in the European Economic Community*, ed. by Charles Steinfield, Johannes Bauer, and Laurence Caby. Newbury Park: Sage.

Drake, William J., and Kalypso Nicolaïdis. 1992. Ideas, Interests and Institutionalization: Trade in Services and the Uruguay Round. In Knowledge, Power and International Policy Coordination, ed. by Peter Haas. *International Organization* 45 (Winter).

Federal Communications Commission (FCC). 1997. In the Matter of Rules and Policies on Foreign Participation in the U.S. Telecommunications Market: Order and Notice of Proposed Rulemaking.

Galbi, Douglas, and Chris Keating. 1996. *Global Communications Alliances: Forms and Characteristics of Emerging Organizations*. Federal Communications Commission, International Bureau.

International Telecommunication Union (ITU). 1996. *Trade Agreements on Telecommunications: Regu-latory Implications*. Briefing Report no. 5 of the International Telecommunication Union Regulatory Colloquium. Geneva: ITU.

International Telecommunication Union (ITU). 1997. *1996/97 World Telecommunication Development Report: Trade in Telecommunications*. Geneva: ITU.

Leebron, David. 1996. In *Fair Trade and Harmonization: Prerequisites for Free Trade?* ed. by Jagdish Bhagwati and Robert Hudec. Cambridge: MIT Press.

Nicolaïdis, Kalypso. 1995. International Trade in Information-Based Services: The Uruguay Round and Beyond. In *The New Information Infrastructure: Strategies for U.S. Policy*, ed. by William J. Drake. New York: Twentieth Century Fund Press.

Noam, Eli M., and Anjali Singhal. 1996. Supra-national Regulation for Supra-national Telecommunications Carriers. *Telecommunications Policy* 20, no. 10 (December).

Petrazzini, Ben. 1996. *Global Telecom Talks: A Trillion Dollar Deal*. POLICY ANALYSES IN INTERNATIONAL ECONOMICS 44. Washington: Institute for International Economics.

Pipe, G. Russell. 1993. *Trade of Telecommunications Services: Implications of a GATT Uruguay Round Agreement for ITU and Member States*. Geneva: ITU.

Rutkowski, Anthony M. 1995. Multilateral Cooperation in Telecommunications: Implications of the Great Transformation. In *The New Information Infrastructure: Strategies for U.S. Policy*, ed. by William J. Drake. New York: Twentieth Century Fund Press.

Tuthill, Lee. 1996. Users' Rights? The Multilateral Rules on Access to Telecommunications. *Telecommunications Policy* 20 (March).

US Trade Representative (USTR). 1997. World Trade Organization Basic Telecommunications Services Talks.

World Trade Organization (WTO). 1994. *The Results of the Uruguay Round of Multilateral Trade Negotiations: The Legal Texts*. Geneva: WTO.

World Trade Organization (WTO). 1997a. The WTO Negotiations on Basic Telecommunications: Informal Summary of Commitments and MFN Exceptions. http://wto.org (March).

World Trade Organization (WTO). 1997b. Data on Telecommunications Market Covered by the WTO Negotiations on Basic Telecommunications. Informal Background Information.

World Trade Organization, Group on Basic Telecommunications. 1997. Communication from the European Communities and their Member States: Schedule on Basic Telecommunications (February).

4

Telecommunications Liberalization: The Goods and Services Connection

JOHN SULLIVAN WILSON

Introduction

The realization of a global information infrastructure (GII) will involve complex changes in technology, industrial structures, and business practices. A true GII will be driven, to a large extent, by the continued convergence of industries in telephone, cable, wireless systems, computers, software, and the video, text, and image providers that supply value-added content. Liberalization efforts through world trade agreements should facilitate this process of convergence, building on a successful series of negotiations over the past 50 years.

US telecommunications leadership, both in advancing domestic deregulation and in forging international agreements, has in fact been a significant factor driving technology convergence and the GII. We do, however, need reaffirmation of consensus on the broad principles that can facilitate the convergence of information technology, telecommunications, consumer electronics, and content technologies with the growth of the internet. These principles should center on competition, deregulation, open and interoperable standards, and recognition that liberalization in services will progress only with understanding of the importance of removing nontariff barriers to trade in telecommunications.

Significant steps have already been made over the past year in removing barriers to trade in converging telecommunications and information

John Sullivan Wilson is vice president for Technology Policy, Information Technology Industry Council, Washington.

technology markets. The two most important examples of this trend are (1) the successes in World Trade Organization (WTO) talks on basic telecommunications services in February 1997 and (2) tariff reductions on information technology goods with the completion of the Information Technology Agreement (ITA) in April 1997.

The next serious obstacle in the realization of the GII is the reduction of nontariff measures such as technical and regulatory barriers to trade in information technology goods. The United States and the European Union (EU) have taken a tentative first step in this direction with the conclusion of the US-EU Mutual Recognition Agreement (MRA) on product testing and certification. We need, however, an aggressive, multipart strategy to remove global nontariff barriers to trade in telecommunications and information technology products. One way to move forward rapidly on removing these barriers—on a multilateral basis—will be the follow-on talks to the ITA, which in Geneva started in the fall of 1997.

This chapter examines principles for future liberalization in information technology and telecommunications markets, with a focus on the linkages between liberalization in services and goods. Special attention is paid to technology convergence and the role convergence may play in new trade talks to remove nontariff, technical barriers. The important lessons learned from the debate and the legal and congressional activities surrounding the implementation of the US Telecommunications Act of 1996 are also outlined. It is likely that key issues in the domestic US debate over introducing competition in local telephone markets will be repeated in the trade debate over liberalization of international networks. Following an overview, efforts to liberalize telecommunications and information technology markets through three key negotiations—the Group on Basic Telecommunications (GBT) leading to the WTO Agreement on Basic Telecommunications Services, the ITA, and MRAs—will be outlined, as well as a recommended framework for the ITA II. The problems in and prospects for liberalizing trade in telecommunications terminal equipment through the recent MRA between the United States and European Union will be analyzed. The conclusion will outline principles that apply to liberalization in both services and goods, as a guide to informing new multilateral trade talks.

Liberalization in US Telecommunications Markets

The US information technology and telecommunications markets provide useful lessons on ways to achieve successful liberalization and market-driven decision making. Deregulation has already enhanced the competitiveness of the long distance industry, bringing significant benefits to both producers and consumers, and the information technology industry is working to achieve the same benefits in local phone service and enhanced telecommunications services.[1] From a producer's standpoint, deregulation has bolstered the US comparative advantage in the telecommunications industry. The Organization for Economic Cooperation and Development (OECD) calculated a US comparative index of 1.7, far exceeding the average index of 1.0 in the 12 large OECD members. The US industry employed approximately 2.3 million workers in 1994. Many worked in highly compensated service-sector jobs. Meanwhile, the absence of regulatory burdens on emerging sectors, such as the internet and enhanced-service providers, has led to low-cost products and services for household and business purchases in the US market and provided a strong foundation for accelerated product growth over the next decade.

The Telecommunications Act of 1996 will be a key part of continued benefits to the US economy through liberalization in service markets. The act was the first major revision of US telecommunications law in 62 years. The fundamental objective of the 1996 act was to eliminate barriers to competition in domestic local and long distance telecommunications markets. (The US country report in the appendix to this volume also discusses the 1996 act in greater detail.) To carry out this broad goal in the act, in a December 1996 speech, Federal Communications Commission (FCC) Chairman Reed Hundt outlined the commission's objectives for 1997: "(1) Make sure all communications markets move from monopoly to competition; (2) secure the public interest in communications; and (3) get rid of all the rules not necessary to reach these other two goals with a red-hot rule burning party" (FCC Chairman Hundt Outlines 1997 FCC Agenda; Pro-competitive, Deregulation Framework Is Goal; Lists Major Issues, FCC Streamlining Efforts, FCC press release, 12 December 1996).

Those skeptical of the benefits of US liberalization in telecommunications have argued that competition will undermine universal, affordable phone service for all Americans. The cost of providing phone service depends mainly on telephone density: it costs more to offer service where houses are located a mile apart than in dense urban areas. Historically, phone companies evened out these cost differences by charging roughly the same rate regardless of density. With complete liberalization, which would drive prices toward costs, rural consumers would pay much more than urban consumers.

Similar examples can be found in the airline industry. Overall, liberalization made US carriers more competitive and lowered airfares. However,

---

1. For an overview of the US Telecommunications Act and the economic benefits associated with deregulation in US markets, see Duesterberg and Gordon (1997). The economic benefits of deregulation in the United States are also addressed in several other chapters in this volume.

some rural areas where few carriers operate are experiencing airfares that are much higher than they were before liberalization. The telecommunications act attempts to balance this kind of effect by driving competition in local and long distance markets while also requiring the FCC and the states to ensure that affordable, quality telecommunications services are available to all consumers. The FCC, in its universal service order, further provided for ways to keep phone service affordable in certain markets—it created subsidies to pay for telecommunications services to schools and libraries in rural and low-income areas.

The 1996 law mandates the removal of legal, operational, and economic barriers that hinder free-market activity by telecommunications firms. Legal barriers are relatively easy to remove. The act of 1996 removes limits as to which firms can offer phone services in the United States. This market-liberalization tool was introduced into the final US offers at the WTO in negotiations leading to the Agreement on Basic Telecommunications Services in 1997. Operational barriers to liberalization, in general, result from technical requirements in service infrastructures and delivery mechanisms. For example, the FCC has eased access to bandwidth through new spectrum-allocation policies in the United States. The commission has also changed network specifications so that consumers can choose alternative long distance companies in the United States without dialing an extra five digits. Economic barriers to liberalization are more complicated and harder to monitor. These barriers must, however, be removed to prevent anticompetitive practices, especially entry-blocking tactics by telecommunications firms.

The following discussion of pro-competitive provisions in the US Telecommunications Act of 1996 illustrates the difficulties of removing economic barriers to the provision of telecommunications services. The lessons from the ongoing implementation of the 1996 act in the United States are clearly relevant for the other 68 countries that agreed to adopt the pro-competitive regulatory principles in the WTO Agreement on Basic Telecommunications Services in 1997.

Procompetitive Principles in Telecommunications Liberalization

The US Telecommunications Act contains ownership caps to foster the creation of a competitive environment; these are largely carried over from prior legislation. For example, the act limits the number of television stations one firm can own to stations that reach 35 percent of all national viewers. The act forbids one company to own two television stations in a single local market or to own both a newspaper and a cable company in the same market. The law also bans the ownership of a cable company and a broadcast company in the same broadcast market. The purpose of these restrictions is to prevent the emergence of a dominant telecommunications company in any single broadcast market. It is feared that the economies of scope and scale enjoyed by such a firm would prevent entry by potential rivals.

Interconnection and Access Charges

The most controversial changes resulting from the 1996 act relate to interconnection terms, access charges, and service charges. Previously, the regional Bell operating companies (RBOCs) had some control over pricing these services through their influence over the state utility boards assigned to regulating interconnection and pricing, although the FCC historically controlled access charges. In August 1996, the FCC issued new interconnection rules to implement the local competition provisions of the act. The rules established a uniform national pricing system for interconnection charges, with a view to fostering competition in local telecommunications markets. The FCC's interconnection order is designed to establish a national framework to stimulate new entrants into local phone markets. Lessons from implementation of the act are therefore relevant to an understanding of how competition in local markets in industrialized countries can be fostered.

The FCC's order has led, however, to intense controversy and debate over implementation of the act. The most contentious part of this debate centers on the FCC interconnection order's pricing provisions. In this debate, the incumbent RBOCs have lined up against the long distance carriers and other potential entrants into local telephone markets. Certain state utility boards and local carriers, including GTE and US West, sued the FCC, arguing that the act does not grant exclusive control over pricing in the local phone loop to the commission. The FCC's new pricing rule would mandate incumbent telephone service providers to price network access according to the commission's estimate of average costs, using the best available technology. RBOC incumbents complained that these rates would not allow them to sustain their networks or recoup embedded costs.

In October 1996, the Eighth Circuit Court of Appeals in St. Louis granted the preliminary injunction requested by local carriers and state boards, delaying the FCC's plan to open local phone markets to new entrants. On 18 July 1997, the court ruled against the FCC, saying the FCC did not have authority to set rules that would have to be followed by state regulators. As the debate between local incumbents and potential entrants continues, the ultimate question must be how the national telecommunications industry can best be positioned to operate under conditions of full competition. The current fragmented regulatory framework at the state level is not ideally suited for future development of this global industry.[2] On the other hand, RBOCs have argued that they should be free, within wide limits, to set interconnection charges and that new entrants should ordinarily be required to build their own networks and not ride free on existing ones.

Apart from the interconnection debate, local telephone companies last year proposed a new pricing system that would have given greater recognition to the volume of telephone usage by local callers. They argued that increased data traffic caused by growing consumer use of the internet over local lines by enhanced-service providers abused the current pricing structure. The RBOCs urged the FCC to impose per-minute carrier access charges, variable with usage, on the providers of enhanced services. A study by Economics and Technology, Inc., commissioned by a coalition of information technology associations and companies, found, however, that fees charged for local telephone usage adequately cover the costs of enhanced service. The FCC decided, in its May 1997 access charge order, not to adopt the telephone companies' recommendations. The debate over pricing and interpretation of the telecommunications act are likely to intensify over the next year, as the FCC completes its rule-making process and as many members of Congress, alarmed by the lack of competition observable since the passage of the act, consider additional legislation.

Universal Service

Another goal of the act of 1996 is to ensure that all Americans have access to new information infrastructures.[3] All US schools, libraries, research facilities, and hospitals are to be connected to the network by the year 2000. To carry out this mandate, in November 1996, the Federal-State Joint Board enumerated the following specific recommendations to accomplish the objectives of the act: (1) availability of quality services at reasonable cost to customers; (2) access to advanced telecommunications services all over the United States; (3) access for all customers, including low-income consumers and rural, insular, and high-cost areas, to telecommunications technologies comparable in quality and cost to those provided to urban locations; (4) an equitable contribution to the

---

2. For further analysis and new FCC rules, see also US country report in the appendix to this volume.

3. Universally accessible services under the law consist of (1) voice-grade access to the public switched network, (2) dual-tone multifrequency signaling or its equivalent, (3) single-party service, (4) access to emergency services (including 911 where available), (5) access to operator services, (6) access to interexchange services, and (7) access to directory assistance. Provisions of the Act are included in the draft FCC Implementation Schedule for the Telecommunications Act of 1996, Public Law 104-104.

advancement of universal service by all telecommunications firms; (5) specific, predictable, and sufficient mechanisms by federal and state government as necessary to achieve universal service; and (6) access for all elementary and secondary schools, health care providers, and libraries to advanced telecommunications services (FCC, Joint Board Adopts Universal Service Recommendations, News Report DC96-100, 7 November 1996).

Historically, the United States, like many countries, promoted universal service through a regulated monopoly that cross-subsidizes poor, remote, and other "deserving" users. On 7 May 1997, the FCC adopted a plan to ensure access to affordable telecommunications services in a more competitive marketplace. The order maintains low phone rates for POTS (plain old telephone service) for consumers through increased rates for second lines and business customers.

There are several serious obstacles to overcome in fully implementing telecommunications liberalization in US markets through the telecommunications act. While the act has provided a strong push toward competition, implementation has proven contentious. The act has, however, provided a strong signal to international liberalization efforts in services at the WTO. The pro-competitive regulatory principles in the act have also served to influence principles for liberalization in goods markets, as they relate especially to trade talks involved with removing regulatory barriers to information technology trade. The next section outlines progress made in liberalization in telecommunications services and also efforts to facilitate trade in information technology goods in a series of important agreements reached in 1997.

Accelerating International Liberalization in Telecommunications Services and Goods Markets

A number of highly significant international trade talks on telecommunications and information technology have occurred in the past four years. The General Agreement on Trade in Services (GATS), a major achievement of talks that led to the creation of the WTO, is the first multilateral, enforceable agreement addressing trade and investment in services. GATS set the stage for important sector-specific negotiations at the WTO, even though the agreement itself simply outlines a framework and statement of guiding principles for liberalization, rather than a set of binding commitments on market access in specific sectors. The overall success of the GATS framework in subsequent sector-specific negotiation will have a substantial impact on US exports. Services account for a significant portion of world trade and nearly 25 percent of US exports. Furthermore, the sector represents a high-growth component of the US economy.

A key accomplishment of the GATS platform is represented in the WTO Negotiations on Basic Telecommunications at WTO, which were concluded in February 1997 after several years of discussions and improved WTO offers.[4] The Agreement on Basic Telecommunications Services was annexed to the Fourth Protocol of the General Agreement on Trade in Services.[5] In total, 69 governments—all OECD nations, 6 Central and Eastern European nations, and over 40 developing countries—agreed on three key components: opening market access, increasing foreign investment opportunities, and adopting a set of pro-competitive regulatory principles (statement of Ambassador Charlene Barshefsky, 15 February 1997).

The WTO agreement, when fully implemented, should significantly reduce the cost of accessing overseas telecommunications markets for international firms. A key part of this cost reduction will depend on the extent to which the 65 governments that agreed to adopt the pro-competitive regulatory principles alter domestic regulatory structures to ensure competition in local markets.[6] In total, signatories to the agreement generate 94 percent of the WTO member countries' total basic telecommunications services revenue, with revenues likely to exceed $600 billion in 1997. As industrial leaders in many areas of telecommunications and information technology, US firms are particularly well positioned to benefit significantly from the agreement.

All other signatories will benefit from the agreement, particularly developing nations and emerging economies, which will be able to access lower-cost, more-advanced telecommunications services worldwide. Improved market-access provisions in the agreement will ensure that telecommunications companies can provide local, long distance, and international service through any network technology (either using their own facilities or reselling the capacity of incumbent carriers). The agreement's provisions on investment opportunities, when fully implemented, will enable firms to establish new telecommunications companies or

---

4. Aronson (chapter 2) and Drake and Noam (chapter 3) discuss the negotiations in detail.

5. For an overview of the GATS and regulatory rules, see Tuthill (forthcoming), Primo Braga (1997), and Hoekman (1996).

6. Signatories include Antigua and Barbuda, Argentina, Australia, Austria, Bangladesh, Belgium, Brunei, Bulgaria, Canada, Chile, Colombia, Côte d'Ivoire, Czech Republic, Denmark, Dominican Republic, El Salvador, Finland, France, Germany, Ghana, Greece, Grenada, Guatemala, Hong Kong, Hungary, Iceland, Indonesia, Ireland, Israel, Italy, Jamaica, Japan, South Korea, Luxembourg, Mexico, the Netherlands, New Zealand, Norway, Papua New Guinea, Peru, Poland, Portugal, Romania, Senegal, Singapore, Slovak Republic, South Africa, Spain, Sri Lanka, Sweden, Switzerland, Thailand, Trinidad and Tobago, the United Kingdom, and the United States. Future signatories include Brazil, Mauritius, Morocco, Turkey, and Venezuela. Partial adoption signatories include Bolivia, India, Malaysia, Pakistan, and the Philippines.

acquire significant shares of existing carriers. Perhaps the most significant part of the agreement was the adoption of a set of pro-competitive regulatory principles based largely on those outlined in the US Telecommunications Act of 1996.

The long-term goals of global competition and liberal markets require the agreement's signatories to adhere to the key "regulatory principles," which are intended to prevent anticompetitive practices in their local markets. The principles are the following (Petrazzini and Kelly 1997):

- Competitive safeguards to ensure that dominant (major) suppliers do not engage in anticompetitive cross-subsidization, do not use information in an anticompetitive manner, and do not withhold essential technical and commercial information.

- Interconnection to ensure that competing service providers can interconnect with the dominant operator (or former monopoly) under nondiscriminatory terms and conditions, at the same cost-oriented and unbundled rates that the dominant operator charges itself or an affiliate and at any technically feasible point. Interconnection procedures, agreements, and rates should be publicly available, and there should be a mechanism to settle disputes arising from interconnection negotiations.

- Universal service, which should be administered in a transparent, nondiscriminatory, and competitively neutral manner and not be overly burdensome.

- Transparency of licensing criteria, in which the terms, conditions, and time required to gain a license should be publicly available. The reasons for the denial of a license should be made known to the applicant.

- Independent regulator, which should be independent of any supplier of basic telecommunications services.

- Allocation and use of scarce resources. Procedures for the allocation and use of scarce resources must be objective, timely, transparent, and nondiscriminatory.

These principles are one important part of the growing linkage between liberalization in goods and services markets in information technology and telecommunications. The principles on services in interconnection, for example, are largely the same as those that might apply to efforts to achieve a reduction in technical regulatory barriers to trade in goods. The provisions on nondiscriminatory interconnection terms for technical standards and specifications overlap with proposals to ensure transparency in national standards systems for goods. The text of the agreement also includes references to the physical characteristics of interconnection and states that the means by which interconnection is provided should be equivalent. One of the principles listed above ensures the right to publicly available licensing criteria, much as the provisions of the WTO Agreement on Technical Barriers to Trade require governments to provide timely notification of new regulations that apply to goods. Moreover, references to a regulatory body with the necessary capabilities to monitor and enforce the provisions in the Telecommunications Services Agreement are similar to those needed to ensure that governments do not use mandatory technical standards to block trade in goods.

Successful implementation of the agreement will require its signatories to clarify its terms, especially in regard to the changes necessary in domestic regulation to support the pro-competitive regulatory principles. As the distinction between technologies blurs, efforts to identify products and services that fall under the agreement grow more complex. The ambiguity of the agreement allowed for a successful conclusion of the negotiations. Successful implementation will now require more explicit guidelines regarding the products and services that fall under the provisions of the agreement. Satellite delivery of video services, for example, is considered a telecommunications service by the United States and is therefore included in the agreement. The European Union considers it to be a broadcast service and thus outside the scope of the agreement. More significantly, the internet may be regarded as an enhanced communications service, a basic telecommunications service, or even a broadcast medium. In the end, the line between broadcasting and other information technologies was not clearly drawn. Technology convergence in telecommunications and information technology, as it continues, will clearly complicate implementation of the agreement. Most importantly perhaps, the way in which the reference paper on regulatory principles is translated into concrete policy proposals represents the key obstacle to successful liberalization.

Trade in Goods: The ITA

Concurrent with the telecommunications services negotiations in 1997, the WTO addressed the increasing significance of trade in information technology products. In March 1997, US Trade Representative (USTR) Charlene Barshefsky announced the successful conclusion of the landmark ITA. This agreement had been championed by US information technology firms and had been brought into discussions with the Quad countries (United States, European Union, Canada, and Japan) and discussions in the US-EU Transatlantic Business Dialogue early in 1995.

The ITA was endorsed by the Asia Pacific Economic Cooperation (APEC) Summit in Subic Bay in November 1996 and at the December 1996 WTO ministerial meeting in Singapore. The final text was outlined and adopted by signatories in early 1997. A total of 28 countries,[7] accounting for 80 percent of world trade, agreed through the ITA that tariffs will be eliminated by January 2000. During talks in Geneva in early 1997, an additional 12 countries pledged their support for the agreement.[8] In total, the signatories at Singapore and Geneva represent 92 percent of the $500 billion-plus global annual information technology trade. The agreement allows for the inclusion of additional members—for example, China and Russia—when circumstances permit. In fact, China notified the WTO in October 1997 that it intends to sign the ITA.

Background of the ITA

The groundwork for an ITA was launched in 1993 when the United States asked the European Union to lower its tariffs on US information technology products. The Information Technology Industry Council (ITI) in the United States had pressed for such tariff reductions in its consultations with the US government. The European Union acknowledged the general economic rationale for liberalization. European firms recognized that they paid a high price because of the European Union's relatively high tariff rates in information technology, including a 14 percent tariff on semiconductors (Tracing Infotech Pact's Roots, *Journal of Commerce*, 13 December 1996). High EU tariffs on semiconductors and other information technology goods had clearly hampered EU competitiveness in global markets.

Progress toward an ITA was made among the Quad countries during their spring 1996 trade conference in Kobe, Japan. At the Quad meeting, the United States pressed for a multilateral ITA to facilitate global trade. It was estimated that liberalization in the sector on top of underlying economic growth would increase global information technology trade to $750 billion by 1999, a 50 percent increase over 1996 levels (WTO Agrees to Abolish Tariffs on Info-Tech Items, *Daily Yomiuri*, 13 December 1996). Following the Kobe meeting, a major step toward the ITA was taken by the APEC Summit leaders meeting in Subic Bay in November 1996. The leaders strongly endorsed the ITA, and in fact, this endorsement became the centerpiece of the Subic Bay declaration. With the groundwork laid in Kobe and Subic Bay, the WTO ministerial meeting in Singapore in December 1996 essentially concluded the negotiations.

---

7. Those countries were the United States, the EU countries (15 members), Canada, Japan, Australia, Hong Kong, Iceland, Indonesia, Norway, Singapore, South Korea, Switzerland, Taiwan, and Turkey.

8. These 12 countries are Costa Rica, Czech Republic, Estonia, India, Israel, Macao, Malaysia, New Zealand, Romania, Slovak Republic, Liechtenstein, and Thailand.

Box 4.1 ITA product coverage: A representative list of products included in the WTO Information Technology Agreement

Computers
- Supercomputers, mainframe computers, workstations, and personal computers
- Automatic teller machines, point-of-sale terminals, calculators, and electronic translators
- All computer peripheral devices, including keyboards, display units, hard disk drives, CD-ROM drives, scanners, digital still cameras, and multimedia upgrade kits
- Local area network (LAN) and other computer network equipment

Telecommunications
- Telephone sets, cordless phones, and video phones
- Cellular phones and pagers
- Telephone answering machines and facsimile machines
- Set-top boxes for internet access
- Telecommunications switching and transmission equipment
- Optical fiber cable

Software
- Application-type and multimedia software products
- Unrecorded floppy disks and other software media

Semiconductors
- All semiconductors, including memory chips, microprocessors, ASICs, etc.
- Semiconductor manufacturing and test equipment

Printed circuit boards
- All printed circuit assemblies for Information Technology Agreement products
- Other passive and active components, including capacitors and resistors
- Smart cards and multichip modules

Source: Press release, US Trade Representative/Geneva, 26 March 1997.

ITA Coverage and Implementation

The ITA signed in March 1997 covers approximately 200 products in the information technology sector, including computers, semiconductors, telecommunications equipment, software, semiconductor manufacturing equipment, digital photocopiers, and capacitors (see box 4.1). While an agreement in principle had already been reached, the Singapore and Geneva meetings featured contentious disputes over the specific products that would be classified as information technology products and that would therefore be eligible for tariff-free status. These disputes reflect, in part, the continued convergence of technologies in information technology markets mentioned earlier in the chapter.

For example, the agreement will cover computer monitors, and digital cameras will be included as they are often linked with computers. The European Union won the battle to include capacitors in the agreement, and the United States agreed to include digital copiers. The United States did exclude some optical fibers and signal electronics. The Japanese successfully included copper cables and chemicals used in semiconductor manufacturing. In the end, the major trading nations of the Quad reached a compromise on product coverage.

According to the agreement, tariffs on information technology products will be reduced in stages, with liberalization commencing in mid-1997. Only the Quad countries, however, have finalized the staging of their tariff cuts on eligible information technology products. All Quad tariffs will be abolished by January 2000. Other signatories are likely to adhere to about the same schedule, but some countries may take longer, until 2005 if necessary, to eliminate tariffs.

According to USTR Charlene Barshefsky, "the movement to zero [tariffs] by the year 2000 in all of these products pretty much constitutes a global tax cut. It means that the creation of the information superhighway will be encouraged and promoted, not taxed." Following the successful conclusion of the WTO negotiations, Ambassador Barshefsky noted that "the significance of the agreement is without comparison. At no time in the history of the trading system have so many countries united to open up trade in a single sector by eliminating duties across the board" (statement at Basic Telecom Negotiations, 15 February 1997).

The ITA: Part II

Building on the momentum of the ITA negotiations, the final agreement included a platform for future discussions on liberalization in the sector. Paragraph 3 of the agreement states: "Participants shall meet periodically under the auspices of the Council on Trade in Goods to review the product coverage specified in the Attachments, with a view to agreeing, by consensus, whether in the light of technological developments, experience in applying the tariff concessions, or changes to the HS nomenclature, the Attachments should be modified to incorporate additional products, and to consult on non-tariff barriers to trade in information technology products."

Work to leverage the ITA using the commitment in final text, commonly referred to as the ITA II, is under way in the US industry through the ITA coalition. The coalition is a group of information technology companies based in the United States and trade associations that supported negotiations leading to the ITA. The coalition is developing proposals to build on the agreement. Under the provisions of the Procedures for Consultations on and Review of Product Coverage, ITA signatories must submit lists of additional products for coverage between 1 October and 31 December 1997. After reviewing the list, the ITA Committee of the WTO will convene no later than 30 June 1998 to decide whether and how to amend the agreement. It is unclear to what extent the ITA signatories and international industry leaders will agree on a final scope and outline of the ITA II.

Within US industry, however, there is strong interest in using the ITA II both to expand the product coverage of the ITA and to address specific nontariff measures that restrict trade in information technology goods. The ITA coalition in the United States is working on a new list of products for inclusion in the ITA scope, possibly including computer batteries, digital video cameras, and video recording devices. A number of nontariff measures are also under consideration for possible inclusion in the ITA-II discussions in Geneva in late 1997. These are forced technology transfer requirements affecting information technology firms and purchasing practices by both government-owned and formerly state-owned companies (statement of the Semiconductor Industry Association on the Review of the Information Technology Agreement, 1997). The global acceptance and use of ATA Carnets, which allow information technology products to travel duty and tax free for use in trade shows and demonstration projects, have also been proposed for inclusion in the ITA-II discussions.

It is clear that the opportunity presented in the ITA II to address nontariff barriers to trade should be taken seriously. They could be a primary focus of the ITA-II negotiations in 1997-98. As tariffs have decreased world wide, manufacturers of information technology products have encountered increasingly serious regulatory barriers to trade. This is particularly true in the areas of duplicative and redundant national requirements on certification relating to electromagnetic compatibility, electrical safety, and attachment to telecommunications networks regulations, among others (ITA-II Recommendations on Technical and Regulatory Barriers to Trade, Information Technology Industry Council, 14 July 1997).

There is a significant opportunity in the ITA II to target the following: (1) duplicative testing, certification, or other technical requirements for information technology products that have already been tested and certified to equivalent standards elsewhere; (2) mandatory accreditation of testing laboratories, including manufacturers' laboratories, that have already been accredited to international guidelines; (3) nontransparency of regulations that affect information technology trade, including certification and labeling requirements, with respect to technical requirements, product coverage, procedures for attesting to compliance, notification, and points of contact; and (4) technical regulations that force disclosure of intellectual property, such as audits, plant inspections, and requirements for detailed technical documentation.

The ITA-II discussions beginning in 1997 could therefore lead to consensus among key ITA signatories on adopting a framework for liberalization in information technology goods that relies on a manufacturer's ability to certify conformity to regulation. The broad goal would be to launch talks leading to a series of global Conformity Assessment Agreements (CAAs) under WTO disciplines. Discussions of these types of agreements in the ITA II could start with a focus on specific regulations affecting information technology goods, including safety, electromagnetic compatibility, and telecommunication network attachment, for example. Individual schedules or roadmaps for a CAA are important, as WTO members have varying amounts of experience with international acceptance of test results and declarations of conformity from suppliers and other laboratories. In addition, regulatory areas covering information technology products, such as electrical safety and electromagnetic compatibility, are closer than others to global acceptance of a single standard.

By 1 June 1998, it should be possible to achieve agreement within the ITA II on at least a framework for talks on the use of global (internationally accepted) standards for meeting regulatory requirements applied to information technology. Signatories to an ITA-II agreement would commit themselves to adopt regulations that dictate global standards in whole or identify equivalency to global standards. This type of agreement would increase the transparency of regulations in information technology trade and eliminate costs associated with demonstrating compliance to multiple, redundant standards.

The ITA-II talks, if successfully launched, could craft implementation schedules in areas where there are currently existing standards with wide international acceptance. For example, agreement could be reached on the use of International Electrotechnical Commission (IEC) 950 standards for electrical safety of information technology equipment. An agreement could also include formal acceptance of International Special Committee on Radio Interference (CISPR) standard 22 for electromagnetic compatibility of information technology equipment. A CAA could also include specific recognition of the acceptance of test results from laboratories that meet the requirements of International Organization for Standardization (ISO)/IEC Guide 25, "General Requirements for the Competence of Calibration and Testing Laboratories." This would greatly facilitate trade in information technology products by removing restrictions on the geographical location of test laboratories. The ITA II could also provide a strong basis for achieving international agreement on the final goal of using a supplier's declaration to attest to conformity to national regulations.

The ITA II offers a unique opportunity to leverage a sector-specific tariff agreement to address nontariff measures. As technologies converge in telecommunications, it is especially important that trade talks address the types of regulatory restrictions on trade that are reflected in protection of services and goods. Removing regulatory barriers on information technology goods is particularly important, as recent outcomes of talks on MRAs to attack technical barriers have proved difficult to conclude on a multisector basis.

MRAs

Bilateral and multilateral MRAs address a more technical and invisible aspect of trade barriers than tariffs: standards. Standards are a vital part of modern commerce. They help assure the safety, reliability, and interconnection of information technology and telecommunications products (see, e.g., David and Steinmueller 1996). Countries with differing technical standards, however, may require that a single product undergo tests in each export market. These duplicative testing and certification procedures, which are meant to ensure compliance with national or international standards, can add significant cost burdens on internationally traded goods and services. It has been proposed that mutual recognition by governments of test data, laboratory competence, and certification requirements, especially in regard to telecommunications terminal equipment, can eliminate unnecessary testing and certification requirements. The MRA model focuses on third-party testing (e.g., testing by independent laboratories), inspection, and certification in sectors regulated by governments through product approval systems. An examination of the recent US-EU talks on an MRA illustrates potential benefits and drawbacks to this model. The conclusion of the US-EU talks also offers insight into the possibilities for agreements in other regions, including an MRA within APEC (Wilson 1995).

MRAs in regulated sectors such as telecommunications offer two potential benefits. Manufacturers would be able to obtain required national certificates at the production location, rather than pay the higher costs of offshore certification. This allows direct shipping of products from point of production to point of final sale, as will hopefully be the case for trade between Europe and the United States when MRAs are implemented. In addition, MRAs could make it possible for products to be tested and certified once, rather than tested and certified in each jurisdiction, significantly reducing obstacles to international trade.

In June 1997, the United States and the European Union concluded an MRA that covers conformity requirements in telecommunications equipment, information technology products, medical devices, and pharmaceuticals and that will specifically address acceptance of test data, laboratory accreditation, and final product certification. In total, the MRAs with the European Union cover over $40 billion of transatlantic trade. The MRA on telecommunications and information technology products alone could, when fully implemented, result in an approximately $1.4

> **Box 4.2 MRA product coverage: A list of products covered by the US-EU Mutual Recognition Agreement**
>
> **Telecommunications terminal equipment**
> - Covers any product "intended for connection to the public telecommunications network in order to send, process, or receive information."
> - Includes analog and digital equipment devices using wired connection (telephones or modems) or radio connection (mobile phones), as well as satellite terminal equipment and radio transmitters.
>
> **Electromagnetic compatibility (EMC)**
> - EMC tests ensure that any equipment does not harm networks or other equipment in the same environment.
> - All tests and certificates issued by agreed-conformity assessment bodies in the territory of one party will be recognized by the other.
>
> **Electrical safety**
> - Electrical safety tests verify the level of risk for workers and consumers of electrical appliances including domestic appliances, hand-held tools, and electrical installation equipment.
>
> *Source*: Information Technology Industry Council, Washington (1997).

billion saving to consumers and manufacturers. Final signature of the MRA is expected by the end of 1997. There will be a two-year, phase-in period, during which there will be mutual acceptance of test data according to US and EU regulations. After the two-year period, certifications performed anywhere by a facility recognized under the MRA in the United States or Europe will be accepted. This should reduce the cost of testing and certification by approximately six to eight weeks' time. Manufacturers will also have a broader choice of testing laboratories. Dispute resolution under the MRA will be handled by a Joint Committee and Joint Sectoral Committee for Information Technology. The MRA between the United States and the European Union will make the most progress in reducing regulatory burdens on trade in telecommunications terminal equipment, although information technology products subject to certain EU requirements will also benefit (see box 4.2).

Lessons of MRA Talks with Europe

There are several lessons from the negotiating experience in the US-EU talks. First, it is clear that approaches to international trade policy in Europe still suffer from political dynamics of the internal market that serve to make external negotiations captive of member governments' protectionist agendas. During the negotiations, for example, the European

Union pressed for the inclusion of rules of origin in the text of the MRAs. US information technology firms, backed by the US government, strongly opposed inclusion of these rules in the agreement. In trade talks on information technology and telecommunications products in particular, it is impossible to determine precise local content. Most importantly, rules of origin retard international trade and do not facilitate it. The European Union abandoned rules of origin demands only in the final weeks of the talks.

The MRA negotiations proved to be extremely time intensive and costly, especially as they were framed to include an overall binding text and sectoral annexes for each product sector under negotiation. The fact that negotiations were undertaken as part of a package, with cross-sectoral tradeoff, impeded the process. The differing regulatory structures in the European Union and the United States also contributed to the difficulties of the negotiations. The US and EU systems differ in both structure and operation, making it difficult to create conditions for exact and reciprocal treatment. While the United States relies primarily on voluntary, self-certification procedures leading to government acceptance of conformity, the EU system centers on third-party conformity assessment in regulated sectors through government-approved facilities.

As noted, the MRA talks included complicated parallel talks on an umbrella text for all the sectoral annexes included in the final package. This probably delayed final completion by a year or more. First, MRAs could probably have been concluded on information technology more quickly as stand-alone agreements. Second, a mutual commitment between national regulatory agencies and trade agencies from the beginning of negotiations is necessary to conclude an MRA. Third, early and high-level political leadership, expressed as a public commitment to a specific date for conclusion of the MRAs, would have significantly accelerated the timetable for completion.

The strong support of industry representatives constituted an important driving force in the conclusion of the MRAs. Throughout the US-EU negotiations, the industry had a strong voice. The Transatlantic Business Dialogue (TABD) in November 1995, for example, demonstrated strong support for US-EU MRA talks from involved industries. TABD also enlisted a commitment for concluding the MRA talks by 1 January 1997. Although the deadline passed without an MRA, it did serve to accelerate the MRA negotiating timetable.

The US-EU MRA will provide lessons for other regions as they address liberalization in telecommunications and information technology. The APEC dialogue has closely examined the US-EU MRA talks. Talks in APEC continue on an MRA on telecommunications equipment, with Canada and the United States pressing for completion of the MRA for announcement at the Vancouver APEC leaders' meeting in November 1997.

Conclusion: Key Principles Relevant to Goods and Services Liberalization

The introduction of new telecommunications services and products, including cellular systems, communications devices, and information technology products, is significantly affected by the degree of regulatory control. In an effort to accelerate the introduction of new technologies, a series of multilateral discussions have taken place to reduce regulatory obstacles to telecommunications trade. Over the past two years, these efforts have focused on three key negotiations that have been designed to pursue the goals of fair competition, globalization, and market-based pricing in the information technology and telecommunications industries. All have ended successfully. These negotiations are only the first step; many obstacles remain to open trade in telecommunications and information technology, especially as the distinction between goods and services in digital technologies continues. The following sections outline some of the guiding principles that will be important as a foundation for new trade talks that link goods and services liberalization.[9]

Rules and Safeguards

The US Telecommunications Act of 1996 serves as a strong incentive for accelerating liberalization of global telecommunications markets. The debate during congressional consideration of the act reflected concerns that deregulation would not automatically lead to open competition. To ensure a competitive outcome, the FCC is in charge of implementing the act of 1996. FCC rulings are closely monitored by countries around the world, many of which are debating the best way to liberalize their domestic telecommunications markets. For many other countries, however, the first step will be the privatization of government-owned monopolies. Privatization alone can easily result in a privately owned monopoly rather than leading to a competitive environment. This is one reason why the pro-competitive regulatory principles in the WTO Agreement on Basic Telecommunication are so critical.

Anticompetitive business practices and technical barriers present a serious nontariff barrier for telecommunications trade in both goods and services. With the reduction of global tariffs through the ITA and the removal of border controls on a wide range of products in the Uruguay Round, anticompetitive business practices have become a serious barrier as import controls, such as tariffs, have been lowered. Without strong national competition policies, private telecommunications monopolies

---

9. For further discussion, see Wilson (1996) and Khemani 1996).

will block market access for foreign competitors and delay the benefits promised by the WTO Agreement on Basic Telecommunications Services. Equally important is the removal of technical and regulatory barriers resulting from product testing and certification, as noted above. In the long term, global agreements such as the CAA outlined in this chapter would prove extremely valuable in expanding trade in information technology goods.

In services, much more needs to be accomplished in outlining the benefits of liberalization through deregulation in preparation for new talks at the WTO. Government and industry must cooperate in educating consumers and producers on the benefits to them in the form of more choices, lower prices, higher profits, enhanced competitiveness, and greater employment opportunities. For example, according to Petrazzini and Clark (1996), the entry of a second cellular operator into the Beijing monopoly market reduced the prices of cellular services by one-third. The same report found that profits of state-owned telecommunications firms in Mexico and Argentina rose dramatically after privatization, while in the markets of China, Chile, and four other developing nations, the introduction of competition led to increased full-time employment in the telecommunications sector. The study concluded, however, that privatization without competition has often driven up prices and reduced employment, demonstrating the need for far-reaching and concurrent liberalization in both areas. In sum, principles that address rules and safeguards to ensure competition will be critical in new rounds of trade talks on liberalizing information technology goods and services.

Interoperability and Open Interfaces

Interoperability in global telecommunications and information technology networks is necessary to enable technology convergence to continue and the trade agreements discussed above to succeed. Thousands of interfaces, both open and closed, exist in today's information infrastructures. A much smaller set of these, located at key high-leverage points, is particularly important. Open interfaces enable the development of new systems, networks, devices, and services that are built and operated by competing providers and users. Conversely, closed interfaces will hinder competition and growth. With closed interfaces, communications providers that possess market power can limit the services that users receive and can restrict their choice of information appliances.

The key to interoperability is the presence of open interfaces among software, communications devices, computers, and other information technology products. Interoperability, in turn, will foster progress toward the GII. An interface in telecommunications networks is open if its specifications are readily available to all vendors, service providers, and users. Technical specifications should be revised only with timely notice in a public process, as referenced in the Telecommunications Services Agreement. Interface specifications—the detailed technical parameters that explain how networks, systems, devices, and services communicate with each other—should be limited to requirements necessary to achieve interoperability. International agreement on common principles that embody the open interconnection and interoperability of voice communications networks must continue to be a part of broad liberalization goals in services. In order to achieve a GII with a full range of services and capabilities, the adoption of effective network interface standards will be critical for other areas as well.

Standards

Standards developed by industry may include both proprietary and non-proprietary technologies. They are critical to the successful launch of global information networks.[10] Where competition exists, the industry-led voluntary standards process, with participation by government as a user, will drive the development of the necessary standards. Where competition is absent and control of a critical interface rests with a company or an individual, national competition laws that ensure open critical interfaces are needed to prepare for the coming era of competition.

Rapid globalization in the telecommunications industry requires that interface standards be developed quickly. Currently, technical barriers to trade embedded in discriminatory certification requirements and regulations that deviate widely from internationally accepted scientific principles present serious problems for firms operating in global markets. These barriers often include testing and certification requirements set at higher standards for imports, costly and discriminatory product-labeling rules, manipulation of domestic laboratory accreditation regimes to block imports, and mandatory compliance with unnecessary quality-system registration schemes. New multilateral talks on regulatory reform standards, such as those in the ITA II, will need to address these barriers.

From the perspective of trade expansion and technology development, a new government role in setting technical regulations in growing digital networks would be extremely damaging. Highly complex and easily manipulatable standards systems controlled by government can quickly affect incentives for investment in research and development and in technology. The work required of manufacturers to achieve and demonstrate conformity to duplicative technical regulations already represents

---

10. See Perspectives on the National Information Infrastructure: Ensuring Interoperability by the Computer Systems Policy Project; Global Information Infrastructure: Industry Recommendations to the G-7 Meeting in Brussels; and GII Interoperability by EUROBIT, JEIDA, ITAC and ITI, 1995.

one of the costliest barriers to international commerce. Opaque national regulations may favor inefficient domestic producers and hinder technological innovation.

In sum, it is difficult to predict the future of rapidly growing and converging industries in information technology and telecommunications. Trade agreements must continually be developed that adapt to innovation and global market structures in dynamic sectors such as information technology. Common market characteristics and business structures that can help inform this process are emerging. Digital content providers and creators, product and services developers, and physical network providers are increasingly exhibiting the following patterns:

- Technological. These research-and-development-intensive industries are all racing to provide state-of-the-art products and services to users.

- Cross-ownership. Mergers, acquisitions, and joint ventures among network providers, content creators, and product and service providers are becoming increasingly commonplace.

- Geographic. Information technology and, increasingly, telecommunications service providers, are now competing in overseas markets as they become liberalized.

- Value-chain position. Each industry sector in goods and services recognizes the value of offering a wide range of services to all customers.

These common features are, in fact, the driving force behind many key public policy issues: domestic and global market-access; agreements on "competitive ground rules"; standards and their link to intellectual property protection; privacy and security debates as electronic commerce grows; and international proceedings on interconnection, interoperability, bandwidth-on-demand, and spectrum allocation. The principles outlined above should be used to shape goals in new liberalization efforts, as business market structures evolve with these common features. The US experience with the Telecommunications Act of 1996, negotiations in the ITA and telecommunications services at the WTO, and the conclusion of the bilateral US-EU MRA offer concrete evidence of the importance of these pro-competitive regulatory principles. They can continue to be relevant in building future trade talks.

References

David, Paul, and W. Edward Steinmueller. 1996. Standards, Trade, and Competition in the Emerging Global Information Infrastructure Environment. *Telecommunications Policy* 20, no. 10.

Duesterberg, Thomas J., and Kenneth Gordon. 1997. *Competition and Deregulation in Telecommunications: The Case for a New Paradigm.* Indianapolis, IN: Hudson Institute.

Hoekman, Bernard. 1996. Assessing the General Agreement on Trade in Services. In *The Uruguay Round and Developing Countries*, ed. by Will Mann and L. Alan Winters. Cambridge, UK: Cambridge University Press.

Khemani, R. Shyam. 1996. The Complementary Relationship between Trade Policy and Competition Policy. In *Regulatory Reform and International Market Openness*. Organization for Economic Cooperation and Development.

Petrazzini, Ben A., and Theodore H. Clark. 1996. Costs of Benefits of Telecommunications Liberalization in Developing Countries. Paper presented at the Institute for International Economics Conference on Liberalizing Telecommunications Services, Washington (29 January).

Petrazzini, Ben A., and Tim Kelly. 1997. *Asia-Pacific and the WTO Agreement: Outcome and Implications, International Telecommunications Union*. Adapted from the World Trade Organization.

Primo Braga, Carlos A. 1997. *Liberalizing Telecommunications and the Role of the World Trade Organization*. Note 120. Washington: World Bank.

Tuthill, Lee. 1996. Users' Rights? The Multilateral Rules on Access to Telecommunications. *Telecommunications Policy* 20, no. 2 (March): 89-99.

Wilson, John S. 1995. *Standards and APEC: An Action Agenda*. POLICY ANALYSES IN INTERNATIONAL ECONOMICS 42. Washington: Institute for International Economics.

Wilson, John S. 1996. Eliminating Barriers to Trade in Telecommunications and Information Technology Goods and Services: Next Steps in Multilateral and Regional Liberalization Efforts. In *Regulatory Reform and International Market Openness*. Organization for Economic Cooperation and Development.

III

DIFFERENT ASPECTS OF THE TELECOMMUNICATIONS INDUSTRY

5

International Trade in Telecommunications Services: An Economic Perspective

HENRY ERGAS

With the successful conclusion of World Trade Organization (WTO) talks in February 1997, globalization of telecommunications is imminent. The WTO Basic Telecommunications Agreement follows a decade of industry-wide transformation driven by both technology and economics. This chapter presents (1) an examination of the cooperative structure and high costs of the existing system; (2) an analysis of the economics of telecommunications against the backdrop of evolving technology; (3) an explanation of the major changes required to create a competitive industry and thereby realize the cost-cutting goals of the WTO agreement; and (4) speculations on future trends of telecommunications trade and investment.

The Cooperative Model

Historically, international telecommunications services have been provided cooperatively by national telecommunications carriers operating on a monopoly basis within their national territory. Cooperation has occurred at three levels: in the provision of service; in the joint supply of major facilities used for international service; and in the setting of technical standards and operating procedures. This cooperative model is largely the result of constraints arising from the regulatory framework prevailing in most countries. For example, regulation has precluded a carrier in any one country from providing "end-to-end" service to customers in another

Henry Ergas is Bell South New Zealand visiting professor of communications and network economics, University of Auckland, New Zealand.

country. In addition, economic factors have buttressed the cooperative model—notably, transactions costs and risk sharing (discussed below).

Cooperation in the Provision of Service

Central to the cooperative pattern for international telecommunications has been the joint provision of service. There are two basic types of service: switched services, such as international direct-dial voice telephony, and dedicated services, such as the provision of leased lines between points in two countries.

In switched services, which are billed at the point where the call originates, the carrier of the originating country buys the service of call termination from the carrier in the terminating country. Conceptually, the terminating carrier accepts the call from a point outside its national territory and transmits the signal to the called party. This transaction, governed by a correspondent agreement between the two carriers, results in the originating carrier's incurring a settlement liability toward the terminating carrier. The amount of this liability is fixed in terms of an accounting rate agreed to by the carriers in their correspondent agreement. The actual settlement rate, payable by the carrier with the most outgoing minutes, is normally set at one-half of the accounting rate multiplied by the number of net outbound minutes.[1] Thus, between two carriers, the carrier sending more calls than it receives incurs a settlement liability on the volume of net traffic. As a result of the settlements process, the carrier in the terminating country indirectly secures a payment from the originating caller, without entering into a contractual relation with that caller and without charging a termination fee to the called party.

In contrast, with dedicated services, the customer generally orders the service at both ends, obtains a "half-circuit" from each carrier, and is billed by both carriers. The actual provision of the circuit involves technical cooperation between the carriers and does not require one carrier to act as the billing agent for the other.

From an economic viewpoint, the distinction between switched and dedicated services made sense. First, consider switched services. Even if the terminating carrier (carrier B) in a switched connection was permitted to enter into a contractual relation with the originating caller in country A, it is likely that carrier B would incur high transactions costs

---

1. This discussion abstracts from the complexities involved in the settlement of third-party charges (e.g., under automated reverse charging arrangements) or operator-assisted services. It also assumes that only two carriers are involved in the transaction; this is clearly inaccurate when the call transits through a third country. When three carriers are involved, the transiting carrier usually secures a share of the settlement payments.

in establishing direct relations and securing payment from the originating caller. Moreover, carrier B would have to ensure that the revenue secured by the originating carrier (carrier A) gave sufficient incentive for A to provide the capacity required to carry the traffic. So the adoption of the settlements process described above is explained by (1) the complexities involved in billing a large number of customers in foreign countries, (2) the difficulties inherent in working out a case-by-case revenue division between two monopoly carriers, and (3) almost universal resistance by called parties to pay for the service of call termination. In these circumstances, the traditional arrangements with a single payment ultimately recovered by the originating carrier from the caller, coupled with a 50-50 division of the accounting rate, worked to reduce transactions costs substantially.

Now, consider dedicated services. These have involved far fewer customers than switched services. Moreover, providing service to each customer has generally involved the earmarking of capacity at both ends of the link for carrying that customer's international telecommunications traffic. Each carrier consequently has to enter into a direct contractual relation with the customer at its end of the link. As a result, it has been fairly easy to effect billing arrangements in which each carrier bills for the service at its end of the link without any financial settlement between the carriers.

Cooperation in the Provision of Facilities

Carriers have also cooperated in providing some of the major facilities required for international telecommunications. Submarine cable systems have generally been laid through a consortia of carriers. Each consortium member covers a share of the costs of the cable system more or less proportional to its expected two-way traffic. Carriers have also jointly funded the two major global satellite systems—INTELSAT and INMARSAT—within the framework of broader intergovernmental agreements. In each case, the facilities have typically been structured as user cooperatives: only the carriers are able to obtain direct service; in return, the carriers incur a financial liability based on actual or anticipated usage.

As with the settlements process, cooperation in the provision of facilities has clearly reflected the constraints imposed by regulatory realities. It is unlikely, for example, that a carrier could have built, owned, and operated a cable system landing in a country outside its service area without violating the exclusive franchise at the foreign end. Moreover, it seems clear that the regulatory constraints have created economic costs. The operation of facilities as user cooperatives entails certain losses in terms of efficient design and management.

Nonetheless, the cooperative structures have had some economic rationale. They spread the risks inherent in lumpy,[2] technically complex, and highly applications-specific investments.[3] Whatever their defects, user-funded cooperatives ensure that at least some part of the risk is allocated to those best placed to evaluate the magnitude and control the extent of it—namely, the carriers at each end of an international link. This allocation of risk was especially important when the technologies involved were relatively unknown, when only the carriers could estimate both the costs entailed and the likely demand for services, and when private capital markets were either excluded from or reluctant to invest in telecommunications-related ventures.

Cooperation in Standards Setting

Carriers have cooperated extensively in setting technical and operating standards for international telecommunications. The subordinate bodies of the International Telecommunications Union (ITU) have been especially important for standardization efforts, in areas ranging from the design of numbering plans (essential for ensuring worldwide addressability) to the signaling systems used for setting up, controlling, and clearing satellite connections. At the same time, the telecommunications regulations of the ITU (which, in contrast to its technical recommendations, are binding on signatories to the ITU Convention) have provided a "standard form" framework for the correspondent agreements that underpin the settlements process. The success of international cooperation in setting technical and operating standards reflects four interrelated factors.

- The first and most obvious is the lack of workable alternatives. In the absence of end-to-end service by a single carrier, communications between countries would not have been feasible without agreement between carriers about their technical and operating standards.

- The fact that the parties to these agreements have not competed directly with one another, and that each carrier has a clear negotiating mandate to speak for customers at its end, has reduced the costs and difficulties of reaching agreement.

- The ability to agree on common technical and operating standards has been assisted by the fact that technological developments, while far-reaching, have been essentially predictable, taken years to imple-

---

2. Lumpy investments are those that are not evenly spread over time. An example of such an investment is one that has a large set-up cost followed by a period of low investment until extra capacity needs to be added.

3. The assets cannot be easily redeployed to other uses. Hence, they have a very low salvage value relative to the original installation cost.

ment, and been controlled by the negotiating parties. The prevalence of a strong engineering culture among the major carriers has also been a significant assisting factor.

■ It made sense to reach agreement on a multilateral basis in order to avoid the costs of negotiating a large number of bilateral agreements, to secure the benefits of global place-to-place connections (which, in economic terms, enlarge the demand side of the market), and to achieve supply-side economies by avoiding stream-specific investments (e.g., dedicated switching systems to interconnect with a single country).

The Rise of Distortions

While the traditional cooperative arrangements made some sense as a means of reducing transactions costs, spreading risks, and ensuring interoperability, they have engendered severe price-cost distortions. These distortions have been most acute in the area of international settlements. All cooperative ventures raise issues about the distribution of the gains from cooperation—what economists refer to as the surplus generated by the cooperative effort. These issues are frequently contentious. The greater the withdrawal of a party's cooperation—and the higher the loss of available surplus that a withdrawal might entail—the larger the share of the surplus that party can seek to claim. These fundamental negotiating realities have led to acute distortions between prices charged for international telecommunications services and the costs of providing those services.

In international telecommunications, the bargaining game surrounding the distribution of the surplus has been broadly bounded by three factors: (1) the requirement to provide international service under the ITU regulations; (2) the formulae set down in the regulations for settlements agreements; and (3), although not binding, the widespread acceptance of the reciprocal, 50-50 split of an agreed on accounting rate as the norm for correspondent agreements. However, these norms have not proven sufficient to avoid serious distributional conflict.

The issues are best understood historically. When the current institutional structure became the norm—during the period from the late 1930s through the 1950s—the agreed on accounting rate on a bilateral link was generally intended to more or less equal the amount that would be charged the calling subscriber at each end of the link. As a result, the settlements process came close to a straight 50-50 revenue sharing, since the rough equality in calling charges generated relatively balanced traffic streams.[4]

---

4. This statement ignores the income elasticity of demand for telephone services. However, these income effects were less significant in the formative period, since most international traffic occurred between industrial countries with approximately the same (high) level of per capita income.

However, the parties to the agreements never bound the amounts they would actually charge customers, an amount usually referred to as the collection rate. Indeed, under the terms of the ITU regulations, collection rates remain under the exclusive control of the carrier at each end. As a result, for a given accounting rate, a carrier can seek to alter its share of the joint revenue by charging its own collection rates, keeping an eye on the collection rate at the foreign end. The difference between collection rates imposed by the two carriers affects the volume of outgoing traffic relative to the volume of incoming traffic. For example, if carrier A charges a high collection rate on international calls and carrier B charges a low collection rate, there will be more traffic from B to A than from A to B. The net flow of the traffic, largely determined by the different collection rates in the two countries, affects the net settlement liability between the two carriers. Hence, the more freedom a carrier has in setting its collection rates, and the greater the discretion it enjoys in limiting international service (for example, by undersizing outgoing trunks at its international exchanges), the greater its ability to enlarge its share of the cooperative surplus. Of course, even a completely unconstrained carrier would not want to push its collection rates to the sky or severely limit outgoing traffic. While these tactics would increase its share of the cooperative surplus, it would simultaneously diminish the size of the surplus—to the carrier's own disadvantage.

As a result of two broad trends, the consequences of revenue manipulation resulting from different collection rates have attracted a great deal of attention in recent years. To begin with, the surplus available for distribution has increased greatly. Beginning in the mid-1960s, rapid technological improvements in international telecommunications sharply cut the costs of providing service. The transition to automatic, customer-initiated dialing in the early 1970s marked a watershed: at one stroke, it reduced the cost of international calls by more than a third and shifted the cost structure from a dominance of operational costs to a dominance of capital costs.[5] Subsequently, the immense reduction in the unit cost of transmission capacity that resulted from the use of fiber optic undersea cable made distance much less significant as a cost determinant—a trend accentuated by the falling cost of satellite capacity, notably in the INTELSAT system.

At the same time, intercontinental traffic volume increased substantially. This increase resulted from rapid technological progress, falling unit costs, lower collection rates, improvements in quality of service, generally rising incomes, and greater integration of the world economy.

In Australia, for example, two-way telephone traffic increased 10-fold in the decade from 1965 to 1974, and then increased some 30-fold over the next 20 years. While expansion elsewhere has been slower, trend

---

5. See the estimates provided in Australian Post Office Commission of Inquiry (1973, 10).

compound annual rates of growth for intercontinental traffic have rarely fallen below 10 percent. The sustained increase in traffic volume has facilitated economies of scale, thus further reducing unit costs. It has also multiplied many times the revenue flows associated with international service, encouraging rival efforts among carriers to garner larger shares of the available surplus.

The positions of parties bargaining over this surplus have become increasingly dissimilar. This is partly the result of shifts in demand. The progressive extension of economic growth to the developing world, the migration of people from developing to advanced economies, and improvements in the telecommunications infrastructure of many developing countries have increased traffic volumes between countries with sharply different income levels. However, the traffic has not expanded in a balanced fashion; in most cases, outbound traffic from advanced economies to developing countries far exceeds traffic in the other direction.

Even more important have been changes in the regulatory environment. In several countries, the provision of international telecommunications services has been opened to competition. These countries account for more than half of outgoing intercontinental traffic minutes; but even so, the bulk of their traffic terminates in countries where monopoly remains the order of the day. As a result, liberalization may increase the share of the producer surplus captured by the monopolist.

This happens because competition drives prices in the liberalized markets down toward costs (including the costs of the net settlement liability). The convergence of prices and costs reduces domestic producer surplus and (assuming relatively elastic demand) increases the volume of outgoing traffic. At a given accounting rate, more outgoing traffic increases the terminating carrier's revenue in absolute terms and its share of the total revenue accruing to both carriers. To be sure, domestic welfare in the liberalized originating market may well rise, because the gains to domestic consumer benefits from lower prices outweigh the shrinkage of producer surplus gathered by the domestic carriers. However, the growing net settlement liability to the foreign carrier is likely to be resented by the domestic carrier, and may equally concern the liberalizing government.

Two features characterize the pattern emerging from these conflicting trends. First, collection rates for international telecommunications remain extremely high relative to the underlying costs of service. The pattern of collection rates for calls to and from Australia provides some insight. Two points are worth making on the basis of the data available: (1) the median foreign-end collection rate is nearly ten times unit costs;[6] and (2) foreign-

---

6. Costs are defined for this purpose as excluding the net settlements receipts of the foreign carrier; including these (as an offset to costs) would make the price-cost gap even greater. Costs are estimated on a stream basis and are defined to include domestic reticulation, international switching, and carriage. All data refer to the Australian network for 1994-95.

Table 5.1 Explaining network unit costs

A. Regression for outgoing unit network costs

R-squared = 0.32

Adjusted R-squared = 0.30

$F_{(2, 78)} = 18.41$

Dependent variable: Outgoing unit network costs [a]

Independent variable	Coefficient	p-value
Ln (traffic density)[b]	-1.16	0.00
Ln (technology level)[c]	-1.14	0.01
Constant	0.25	0.09

B. Regression for incoming unit network costs

R-squared = 0.31

Adjusted R-squared = 0.29

$F_{(2, 78)} = 17.50$

Dependent variable: incoming unit network costs [a]

Independent variable	Coefficient	p-value
Ln (traffic density)[b]	-1.54	0.00
Ln (technology level)[c]	-1.11	0.01
Constant	-0.01	0.00

Note: All variables have been transformed so as to protect the confidentiality of the underlying data. The symbol Ln indicates logarithm to the base e.

a. Unit costs are estimated on a stream-by-stream basis and include domestic reticulation and international switching and carriage, but exclude net settlement payments or receipts.

b. Traffic density is measured as the number of outgoing minutes for each stream.

c. Technology level is proxied by the number of main lines per 100 population.

Source: Author's calculations.

end collection rates tend to be highest relative to Australian outgoing charges in countries that are relatively distant from Australia and in countries that have no competition in the provision of international service.[7] These results reflect the importance of competition in determining price levels. In addition, the distance-dependence of foreign-end collection rates contrasts sharply with the structure of costs. Analysis of stream-by-stream unit costs (excluding settlements) for Australian traffic suggests that distance plays a small role in cost determination. This is borne out by the estimates presented in table 5.1, which shows that unit costs depend

---

7. The simple correlation coefficient is -0.27 between an indicator of the extent of competition at the foreign end and a dummy variable that takes on a value of one if the ratio of the foreign-end collection rates to the Australian outgoing collection rate is in excess of the median value for this ratio.

primarily on traffic density and on the technology level at the foreign end (a variable proxied by the number of main lines per 100 residents); distance was not a statistically significant variable.

Second, high collection rates are associated with accounting rates, which are extremely high relative to long-term incremental costs. Again using Australian data, the median nominal accounting rate (the outpayment per minute of total outgoing traffic) is some five times network unit costs; the median effective accounting rate (the outpayment per minute of net outgoing traffic) is some 1.4 times network unit costs. A quarter of effective accounting rates are four or more times network unit costs. Moreover, the higher the accounting rates, the higher foreign-end collection rates tend to be, both in absolute terms and relative to Australian collection rates (which are constrained by competition).[8] A 1 percent increase in the nominal accounting rate is associated with a 0.7 percent increase in the foreign-end collection rate and only a 0.5 percent increase in the Australian collection rate;[9] a 1 percent increase in the effective accounting rate is associated with a 0.2 percent increase in the foreign-end collection rate and a less than 0.1 percent increase in the Australian collection rate. In monopoly markets, nominal and effective accounting rates, together with collection rates, are best viewed as being set jointly. The data suggest that carriers that set high nominal accounting rates also set high collection rates, thereby securing a high effective accounting rate by reducing outbound traffic.[10]

Patterns of Distortion

But it is misleading to view these price-cost distortions mainly as attributes of economic development. To be sure, at a superficial level, there appears to be an association between per capita GDP on the one hand, and nominal and effective accounting rates on the other (see table 5.2). Taken at face value, the results in table 5.2 suggest that a 1 percent increase in GDP per capita reduces the nominal accounting rate by close

---

8. Analysis of the impact of accounting rates on collection rates faces obvious problems of identification and simultaneity bias. Negotiators set accounting rates with some expectation of the post-negotiation price-setting behavior of the foreign carrier. The estimates presented in the text need to be read with this in mind. In contrast, the structural variables discussed below can be taken as fully exogenous; and so long as the negotiating game is sub-game perfect, the reduced form equations presented later are identified as to cause and effect.

9. Foreign-end collection rates are in any case substantially higher on average than Australian outgoing charges.

10. Holding per capita GDP constant, a 1 percent increase in the ratio of the foreign collection rate to the Australian outgoing charge reduces the ratio of inbound to outbound switched minutes by -0.35 percent.

Table 5.2 Determinants of nominal and effective accounting rates: The role of per capita GDP

A. Regression for nominal accounting rates

R-squared = 0.51

Adjusted R-squared = 0.46

$F(3, 32) = 11.06$

Dependent variable: nominal accounting rates[a]

Independent variable	Coefficient	t-value
Ln(per capita GDP)[b]	-1.95	-2.91
Ln(competition)[c]	-0.57	-2.29
Private[d]	-1.09	-1.65
Constant	10.68	4.43

B. Regression for effective accounting rates

R-squared = 0.31

Adjusted R-squared = 0.25

F-test $(2, 24) = 5.44$

Dependent variable: effective accounting rates[e]

Independent variable	Coefficient	t-value
Ln(per capita GDP)[b]	-0.63	-2.34
Ln(competition)[c]	-0.12	-1.21
Constant	1.83	1.87

Note: All variables have been transformed so as to protect the confidentiality of the underlying data. The symbol Ln indicates logarithm to the base e.

a. Nominal accounting rate is the outpayment per minute of outgoing traffic.

b. Per capita GDP is measured in US dollars, translated at nominal exchange rates, for the year 1994-95.

c. Competition is a variable containing information on the extent of competition in each country. It is based on whether facilities-based entry was allowed, only resale was allowed or tolerated, competing supply of value added services was allowed, or none of the above.

d. Private is a dummy variable that takes a value one when the telecom industry is privatized and takes a value zero otherwise.

e. Effective accounting rate is the outpayment per minute of net outgoing traffic.

Source: Author's calculations.

to 2.0 percent and the effective accounting rate by 0.6 percent. (Parenthetically, it is worth noting the significant impact of the competition variable in determining nominal accounting rates. Private versus public ownership also helps explain the pattern of nominal accounting rates, but to a lesser degree.)

However, these superficial results need to be interpreted with caution. In particular, the seeming ability of economies with lower levels of per capita GDP to secure higher nominal and effective accounting rates needs to be explained. Two sharply contrasting explanations are typically given by industry participants.

The first stresses the investment requirements of poorer countries and asserts that higher accounting rates are intended to cover the costs of expanding the domestic telecommunications infrastructure. Although simultaneity makes this a difficult model to test, it does not sit comfortably with the data. In particular, holding other structural factors constant, investment per main line is negatively related to the level of nominal and effective accounting rates. In other words, countries that are expanding their network most rapidly tend to have significantly lower (rather than higher) nominal and effective accounting rates. This result appears inconsistent with the "investment needs" explanation.

A second, more compelling, explanation emphasizes the "cash cow" aspect of international telecommunications. In this explanation, some countries view high accounting rates as a way of imposing a hidden tax on domestic and foreign consumers. The fact that this tax is paid in foreign currency makes it all the more valuable in countries with misaligned exchange rates. This in turn suggests that nominal and effective accounting rates are likely to be highest in a country where (1) domestic consumer preferences receive little weight (so that carriers can impose relatively high collection rates), and (2) market distortions increase the value of foreign currency earnings relative to their value at the official exchange rate.

This hypothesis receives considerable support from the data. Thus, table 5.3 sets out a model of the nominal and effective accounting rates that includes three variables obtained from a recent study of patterns of economic development: a variable capturing extreme political instability; a variable capturing political repression; and a trade regime dummy variable that takes a value of 1 for countries that have open trade regimes (Sachs and Warner 1995). Together with the competition variables, these indicators do a far better job than per capita GDP in explaining the pattern of nominal and effective accounting rates, as seen by comparing the adjusted R-squared values in tables 5.2 and 5.3.

In short, the pattern of nominal and effective accounting rates appears to be more significantly influenced by a country's political environment and broad economic policy choices than by its level of development. The highest accounting (and collection) rates are typically imposed in countries that have repressive political systems and closed markets.

Toward Competitive Globalization

The distortions associated with high accounting and collection rates impose substantial economic costs; while these distortions have to some

Table 5.3 Determinants of nominal and effective accounting rates: The role of structural variables

A. Regression for nominal accounting rates

R-squared = 0.78

Adj. R-squared = 0.74

F-test (4.27) = 3.28

Dependent variable: nominal accounting rates[a]

Independent variable	Coefficient	t-value
Political instability[b]	0.30	3.01
Political repression[c]	0.92	6.11
Competition[d]	-0.00	-3.18
Trade dummy[e]	-0.76	-1.07
Constant	0.94	13.21

B. Regression for effective accounting rate

R-squared = 0.50

Adj. R-squared = 0.41

F-test (3.18) = 5.88

Dependent variable: effective accounting rates[f]

Independent variable	Coefficient	t-value
Political instability[b]	0.87	2.19
Political repression[c]	0.67	1.00
Competition[d]	-0.20	-2.21
Constant	-0.15	-0.21

Note: All variables have been transformed so as to protect the confidentiality of the underlying data.

a. Nominal accounting rate is the outpayment per minute of outgoing traffic.
b. The value of political instability is larger for a country with an instable political situation, and smaller for a country with a stable political situation. Please refer to Sachs and Warner (1995).
c. The value of political regime is larger for a country with a more repressive political nature, and smaller for a country with less repressive political nature. Please refer to Sachs and Warner (1995).
d. Competition is a variable containing information on the extent of competition in each country. It is based on whether facilities-based entry was allowed, only resale was allowed or tolerated, competing supply of value added services was allowed, or none of the above.
e. Trade dummy takes a value of one for countries that have open trade regimes and a value of zero otherwise.
f. Effective accounting rate is the outpayment per minute of net outgoing traffic.

extent been undermined by market forces, the preference of monopoly carriers to restrict international service means that distortions are not on the edge of extinction. However, the WTO Basic Telecommunications Agreement will accelerate the transition from monopoly to competitive markets and help relegate huge price-cost distortions to the archives of economic history. In anticipation of these changes, three points need to be made.

First, demand for international telecommunications is relatively price elastic. Even in Australia, where collection rates are relatively low, the price elasticity of demand for switched outbound minutes is significantly above 1.0. This implies that, when a collection rate increases by 10 percent, the volume of international calls decreases by more than 10 percent. For outgoing call streams on which accounting rates are highest (and, hence, collection rates are also high), the price elasticity of demand lies in a range near 1.5. In this case, a 10 percent increase in collection rate leads to a drop of about 15 percent in the volume of international calls. At the same time, long-term incremental costs, excluding net settlement payments, are extremely low. Thus, it is the effective accounting rate that essentially constrains price reductions by Australian carriers. With capacity abundant, the foregone consumer benefits represent pure waste. Demand elasticities are certainly well above Australian levels at the foreign end of many of the call streams with the highest accounting (and collection) rates. As a result, the combined effect in Australia and abroad must amount to a considerable loss of social welfare (using the economist's yardstick of the sum of consumer and producer surplus). Future transmission and switching systems have even lower long-term incremental costs. Therefore, the burden of existing distortions can only increase if collection rates remain high.

Second, on the outgoing call streams with competition at both ends, market forces are bringing accounting rates down; they are likely to bring collection rates down even more rapidly as competitive rivalry intensifies. In part, this simply reflects the keen interest of competitive carriers in reducing their input costs. But several additional factors are also at work:

- As regulatory constraints are lifted, accounting rates are increasingly bypassed by carriers operating on an end-to-end basis. In some cases, these carriers own facilities at both ends of an international connection (for example, Sprint on the transatlantic and now transpacific routes); in others, they lease capacity from a facilities-based carrier and resupply international services over that capacity (for example, Esprit, ACC, and WorldCom on the UK-US route). Strong competition in the supply of international capacity keeps the lease rates paid by new entrants down; as a result, new carriers can substantially undercut the collection rates set by incumbents operating within the framework of the international settlement system. In London, for example, the average collection rate per minute of switched traffic to the United States charged by resellers has been about one-third lower than the rates charged to corporate customers by British Telecom.

- New entrants have exploited anomalies in the structure of accounting rates and international charges. Accounting rates have typically been set on a relatively discriminatory basis: a carrier in country A may charge a carrier in country B a higher rate to terminate its traffic than it charges a carrier in country C, even though the costs entailed by traffic from B are no higher than the costs entailed by traffic from C. Similarly, the A carrier may charge consumers more for calls to B than to C, even though the costs to call C are no higher than to call B. This results in manifold anomalies, including substantial differences in charges in the two directions of a stream; gains from triangulation (it may be cheaper to call C indirectly from A via B, rather than to call C directly from A); and scope for accounting rate arbitrage (a new entrant in A can save money by transiting traffic to C via B, claiming to carrier C that the traffic it has received originates from B not A—a practice known as refile). Digital technology, which allows complex routings and charge reversals without substantial additional costs or any loss of service quality, makes these anomalies easy to exploit.

- New telecom entrants and consumers are proving adept at exploiting emerging opportunities for substitution between services—including the transfer of traffic between public switched networks, private virtual networks, and fully dedicated networks. The rapid growth of email over the internet must be causing loss of traffic to other services and uses, especially fax. With internet access prices set around $1 per hour (as against average collection charges for international direct dial still in excess of $1 per minute), the potential impact of voice transmission via the internet is far-reaching. For example, the number of commercial internet customers in Australia is increasing by some 8.5 percent per month. Internet capacity to the United States from Australia expanded by more than seven times between 1994 and 1996.[11] As existing frame relay networks are upgraded to cell relay, they too will pose a very substantial threat to the existing switched services.

- In combination, these factors are forcing the incumbent carriers to respond, notably by positioning themselves for substantially lower collection rates. This, in turn, is leading to renewed efforts by these carriers to bring costs under control, including a cut in the termination charges they pay to foreign carriers.

---

11. For a comprehensive survey, see Strategic Innovation Group (1995, 33).

The third point about antidistortion pressures is that they cannot completely counteract foreign-end market power. The pressures are more effective in eliminating anomalies in the structure of accounting and collection rates than in forcing down the average level of those rates. Several illustrations suggest how monopoly carriers can respond to the pressures discussed above:

- Consider, for example, callback operations, which exploit the difference between collection rates in the two directions of a stream. The simplest way for a foreign-end carrier to protect itself against callback is to post an increased accounting rate, since this ensures that the losses it incurs in foregone collections are offset by revenues from additional settlements payments. The result is to raise the equilibrium price in the competitive country rather than to lower the price in the monopoly county. What is at issue here is not the absolute level of accounting rates but their level relative to costs (which are falling rapidly). Therefore, a monopoly carrier can protect itself from callback operations simply by slowing the pace at which accounting rates are reduced.

- Refile may also make it difficult to sustain anomalies in the triangular patterns of accounting and collection rates and termination charges, and will thus likely help eliminate discrimination. But the more uniform rate structure will not necessarily be lower than the structure now in place. Indeed, it could be higher if higher rates enlarge the revenue streams accruing to monopoly carriers. Equally, carriers with monopoly power can guard against interservice arbitrage by eliminating anomalies in their termination charging structures—for example, as between switched and dedicated capacity.

- Finally, without effective regulatory constraints, carriers from monopoly markets can increase their accounting rates by taking advantage of bypass opportunities in markets open to competition. In Australia, for example, one foreign carrier whose home market is protected from competition has leased capacity from its home base into Australia, where it will deliver traffic to an out-WATS service, incurring termination charges substantially below the accounting rate. As a result, the foreign carrier will escape settlements payments, while Australian carriers will have to pay the established accounting rate on a much higher level of imbalance minutes.

Overall, liberalization doubtless provides substantial benefits to consumers in the markets being opened to competition. But it will not eliminate, and, indeed, it may even increase, the distortions arising from the conduct of monopoly carriers. These realities help explain why competitive countries pressed so hard to conclude the WTO Agreement on

Basic Telecommunications. They also explain why the US Federal Communications Commission has taken such a hard line—independent of the WTO Agreement—on accounting rate reform.

Trends and Prospects

Looking to the future, it seems clear that the traditional cooperative arrangements that have characterized international telecommunications are giving way to new forms of globalization; what is fundamentally at issue is whether the mechanisms forged in the past can adapt to the new realities.

The most obvious factor underpinning these new realities is the opening of markets to competition. The regulatory constraints that previously forced cooperative provision are being steadily removed; the liberalization of telecom markets in the European Union represents a further, very major step in this direction. Assuming that barriers to foreign direct investment in the telecom sector continue to fall, regulation will no longer be such a potent factor impelling carriers to cooperate in providing service, building and operating facilities, and setting technical standards.

At the same time, many of the transactions cost efficiencies associated with the traditional forms of cooperation may be waning. Consider, for example, the old norm of joint telephone service through correspondent relations. This was partly justified, as discussed earlier in this chapter, by a substantial saving of transactions costs. However, with new billing technology and caller ID numbers, the costs that a foreign-end carrier now incurs to provide service on an end-to-end basis are far smaller. Moreover, by expanding its customer base, the foreign-end carrier can exploit opportunities to sell a broad range of services into the host market (e.g., credit cards). It can also secure comprehensive utilization of its home country facilities at times when they would otherwise be lightly loaded. For instance, the carrier can use its switching system in the home market to control calls in a country located in a different time zone and, hence, with a noncoincident traffic peak. The very high quality of today's transmission and signaling network allows complex call routings with no loss of quality, and this means that facilities no longer need to be located near the point of primary use.

Similar factors are at work to undermine cooperation in the joint provision of facilities. The argument that risk sharing justifies retaining the current structure of INTELSAT and INMARSAT seems less compelling when faced with the private sector's demonstrated willingness to invest in ventures such as PANAMSAT and the next wave of Low Earth Orbit global satellite systems. Rather, the issue is how the strengths of the cooperative systems can be preserved in a more competitive environment. While this is a matter of lively debate—most notably with respect to the new INMARSAT commercial subsidiary and the proposed re-structuring of INTELSAT—it seems likely that the traditional coopera-tive arrangements will play a smaller role in providing the facilities of the future.

Finally, the rise of end-to-end services, the ever more competitive na-ture of relations between carriers, and the increasing pace and com-plexity of technical developments are all undermining the traditional cooperative approach in setting standards. There are strong arguments for interoperability; but even with recent reforms, the ITU mechanisms for ensuring interoperability face substantial difficulties. As a result, de facto and proprietary standards—some of which may be technically open (in the sense of allowing transparent interconnections)—are likely to be-come ever more important.

Many of these changes are already well under way. Forging global alliances between carriers is a symptom of the pace at which develop-ments are unfolding. But these changes do not mean that correspon-dent relations will disappear; rather, they are evolving in two directions.

First, in some instances, correspondent agreements are moving from their traditional form to something closer to a joint venture. Compared to their predecessors, these new agreements involve cooperation on a broader set of issues and allow a wider range of mechanisms for shar-ing revenues and risks.

Second, there are many cases where it is not worthwhile for a foreign carrier to become closely and directly involved in the home market. All it really needs is a set of services close to those traditionally provided in the correspondent relation. In these cases, home carriers increasingly compete to provide the desired services on a commercial basis. One important result of competition is the progressive unbundling of the outbound, inbound, transit, and refile functions into separate contrac-tual elements.

The net effect of these changes is that international telecommunications services trade, which traditionally reflected accidents of geography and the impact of regulation, will be increasingly determined by underlying comparative advantage—most notably by the ability of service providers to generate innovative, cost-effective services. While the transition will take some time, the opportunities are apparent.

Standing apart from this wide array of changes are the countries that remain committed to the old way of doing things. Reluctance to open their markets to competition, and the persistence of high accounting and collection rates insulate these countries from new developments. Many of the national schedules submitted in the WTO agreement seem designed to preserve the old way of doing business. This strategy may reap some short-term rewards, but its long-term costs are likely to be considerable.

Most importantly, the countries in question are likely to be isolated from emerging technology. With relatively small traffic volumes, and constraints on the range of service providers, there may be little scope for the introduction of new services. There is already evidence that countries with monopoly carriers are lagging behind. For example, after adjusting for per capita GDP, countries with network competition are 13 times more likely to offer international Integrated Services Digital Network (ISDN) than others. This is one small indication of the high cost of monopoly carriers.

References

Australian Post Office Commission of Inquiry. 1973. *Supplementary Material Submitted by the Post-Master General's Department* 13. (August): 10.

Sachs, J., and A. Warner. 1995. Economic Regulation and the Process of Global Integration. *Brookings Papers on Economic Activity* 1: 1-118.

Strategic Innovation Group. 1995. *Internet Access Providers in Australia* (December): 33.

6

Telecom Mergers and Joint Ventures in an Era of Liberalization

ROBERT W. CRANDALL

The rate of technical change in telecommunications is causing major upheavals in national telecom policies. Twenty years ago every country's telecommunications sector was a state-controlled monopoly, and in most it was even a state-*owned* monopoly. Apologists may claim that this state of affairs was required because telecommunications services were produced under conditions of natural monopoly. In fact, no one knew whether this monopoly was "natural" or not, and most national telecom authorities (PTTs) surely did not want to ask the question.

Today, the genie is out of the bottle. Competition—or at least liberalized entry—is replacing government-protected monopoly in most developed countries and even in many developing countries. State-owned monopolies are being privatized; foreign capital is being invested in these erstwhile monoliths or in competing networks; joint ventures among large telecommunications and media firms are proliferating; and new services are being developed with little assurance that there is a market for them.

Of course, these changes are not without cost. Doomsayers are quick to point to the loss of jobs in the older enterprises, jobs that were quite attractive given that the employment agreements were reached without the nettlesome effects of competition. Officials responsible for competition policy worry that the erstwhile monopolists will never permit full competition because they control essential "bottleneck" facilities—the local connections to the country's telephone subscribers. On the other

Robert W. Crandall is senior fellow at the Brookings Institution, Washington.

hand, many telecom authorities are now beginning to fear that the new entrants may succeed so well that the old telephone companies may be forced to write off billions of obsolete plant. Nationalists fear that international competition and multinational telephone companies may place their countries' essential communications infrastructure in the hands of aliens. Finally, on a more mundane level, businesses and consumers must now make their decisions from an expanding array of confusing choices among long distance carriers, terminal equipment, voice-messaging services, paging services, personal communications networks, or cellular services.

In this paper, I provide a survey of some of these market developments, particularly those involving mergers and joint ventures among telecommunications firms, and I offer some analysis of their likely threat to the development of competitive markets. Given the dizzying pace of technical change in this sector, no one can predict how telecom markets will develop 5 or 10 years from now, but I shall at least hazard a few educated guesses.

The End of the Dinosaurs

National telephone monopolies have been the rule for a century—100 years in which technology has evolved from the primitive equipment used to implement Mr. Bell's (and others') patents to a set of facilities that reflect much of today's electronics revolution. Nevertheless, telephone service developed in virtually every country as a service provided by the state and therefore subject to severe political constraints. The service was delivered by adjuncts of the post office, surely not prime candidates for efficient organization. Service levels and rates were heavily influenced by politics. For instance, in most developed countries, rural areas received telephone connections at rates far below costs. Long distance rates were correspondingly far above costs. And international rates could only be described as virtually extortionate. (For an analysis of international rates, see Johnson [1989]; Organization for Economic Cooperation and Development [OECD] [1994].) In fact, in virtually every country, many of these pricing distortions continued to this day. Telephone rates are not reflective of the relative costs of service in any country today, not even those that have liberalized the most, such as the United Kingdom, Canada, and the United States. (For an analysis of rates and costs of telephone service in the United States and Canada, see Crandall and Waverman [1996]; for a comparison of rates across countries, see OECD [1990]).

Competitive entry might have occurred decades ago but for state ownership of the telephone system (or, as in the case of Canada and the United States, but for regulation). However, it is not clear that such entry would have been successful when the technology was little more than copper wires and electromechanical switches. It was only when technology began to change after World War II that entry became a serious possibility. By 1959, the US federal regulator, the Federal Communications Commission (FCC), moved to allow private nontelephone firms to build their own (microwave) communications links. By the early 1970s, the FCC had actually licensed competitive (long distance) common carriers in the United States, but other countries would not follow until the mid- or late 1980s (Crandall and Waverman 1996, chapter 1).

By the end of the 1980s, telecommunications technology had changed so dramatically—even from its 1970s dimensions—that the embedded plant and service offerings of most national telecommunications carriers were being threatened by obsolescence. Fiber optics, computer-like digital electronic switches, highly sophisticated PBXs, and other customer equipment were now state of the art, and the human-capital requirements for telephone company employment had changed dramatically. But without competition to spur the diffusion of this new technology, national telephone companies failed to adapt completely to the new order. Pressures from the business community for reliable, low-cost telecommunications provided the force for liberalization and privatization in many countries, such as Canada and the United Kingdom. In the United States, however, an antitrust suit brought in 1974 provided enormous new impetus for the liberalization that had begun anyway. AT&T was broken up in 1984. Japan and the United Kingdom followed by privatizing their national carriers in 1984-85 and allowing at least limited competitive entry. Somewhat later, New Zealand privatized its national carrier and then totally deregulated the telecommunications sector. Australia admitted a second carrier, Optus, and the European Union is actively pursuing a massive liberalization that is due to begin in 1998.

These changes have exposed policymakers and their national telephone companies to enormous pressures. Communications-intensive businesses, particularly in the financial sector, cannot compete if they must rely on expensive, outdated telecommunications services, but national governments are often reluctant to allow liberalization for fear of exposing the employees of the national carrier and its equipment suppliers to massive layoffs and subjecting subsidized residential ratepayers to market realities. Nevertheless, in developed countries at least, there would appear to be no alternative to liberalization because of the essential nature of telecommunications in the postindustrial world.

New Technologies and New Services

The rapid pace of technical change in telecommunications is not simply reflected in lower-cost and more reliable voice-data services that we once obtained through a black rotary telephone connected to copper wires and, perhaps, an acoustically coupled modem. Rather, the new technologies now allow an incredible range of sophisticated, high-speed services. Local area networks, which connect desktop computers and servers, are proliferating. Digital transmission speeds are pressing to levels of 150 megabits per second and upwards. A pair of copper wires can now deliver a video signal. A coaxial cable can deliver 60 or more such signals and may soon be able to offer several hundred. Direct-broadcast satellites are capable of 200 such channels today. Portable phones have become so small and inexpensive that virtually anyone can afford to carry one around in a pocket or purse. Orbiting satellites will soon be able to provide ubiquitous voice-data service over the entire globe. The possibilities are seemingly endless.

These changes may sound like good news to the casual observer, but to incumbent telephone companies, they offer both new opportunities and an increased risk of disaster. Soon, in most developed countries there will be four, five, or more cellular-like telephone services. Most will not be owned by the established terrestrial telephone carrier. Cable television companies in the United Kingdom are offering telephone service, while cable companies in the United States, Canada, and Australia are currently upgrading their plants to offer telephony. Private networks have proliferated in the United States and are likely to grow in other developed countries. The telephone companies' traditional markets are being eroded by competition, but these old, traditional companies are far from obvious candidates to venture forth into the brave new worlds of distributive-video or switched-video applications. After all, it was not AT&T or British Telecom that developed DirecTV, the internet, or CNN. If national governments can no longer be relied upon to protect the erstwhile national carriers, what are these carriers to do?

To cope with these market threats and opportunities, the older telephone companies are currently using three major strategies. First, the companies are attempting to defend their most important sources of cash flow, such as overpriced international services. Second, the more dynamic of them are attempting to exploit their technical expertise in telephony by investing in traditional telephone companies in other (newly liberalized) markets. Finally, they are desperately trying to position themselves to deliver video services. Whether any of these strategies is generally successful remains to be seen, but one can understand the wave of joint ventures and foreign investments as part of one or more of these strategies.

The International Telephone Services Market

It is now possible to send voice signals to virtually any part of the world at peak calling hours at an incremental cost of no more than 12 cents

Table 6.1 Major joint ventures among large international carriers

Participants	Type of venture
AT&T and several others	Marketing arrangement
British Telecom/MCI	Joint venture; British Telecom purchases 20 percent of MCI
British Telecom/MCI	BT purchases remaining 80 percent of MCI
France Telecom and Deutsche Telekom/Sprint	Joint venture; France Telecom and Deutsche Telekom purchase 20 percent of Sprint
AT&T, Singapore Telecom, KDD (Japan)	Worldsource, an Asian Alliance

per minute and often much less (FCC 1996, 51). Yet most international telephone rates are above $1 per minute, often substantially above. This price-cost situation has led to incredibly complicated arbitrage systems that allow some large international users to obtain rates of perhaps 14 cents per minute while individual consumers use a major national carrier to provide the same service at more than 10 times this price. Clearly, the large national telephone companies want to try to stop such arbitrage or at least to limit its effect through highly discriminatory tariffs. Joint ventures among telephone companies, such as those listed in table 6.1, are designed to do a little of each.

Because most countries, including the United Kingdom, Canada, France, and Germany, have not opened their international services markets to foreign entry, these ventures are extremely controversial. Moreover, given the enormous rates that are charged for terminating international traffic, such ventures are generally seen, particularly by some US carriers, as clandestine attempts by the merger partners to divert terminating traffic from their rivals to their own networks. The irony of the current international pricing system is that countries with high outbound rates that are due to an absence of competition trigger elaborate call-back procedures that reverse the flow of traffic, substituting lower-priced inbound calls for their overpriced outbound calls. But these procedures add to the imbalance of traffic, increasing the flow of settlement charges for terminating calls in the restrictive country. Thus, the "penalty" for charging high international rates is that the foreign monopolist receives a disproportionate share of high-priced termination fees for passively accepting its foreign counterparts' calls.

These international joint ventures—for example, British Telecom/MCI or Sprint/France Telecom/Deutsche Telekom—provide an opportunity for the liberalizing countries, such as the United States, to disallow foreign investments in their own carriers unless the foreign companies' markets are also opened to competitive entry or foreign investment. Unfortunately, such negotiation strategies may not bear much fruit, a fact now virtually conceded by US trade authorities.

Recently, the US FCC has moved to force major reductions in international accounting rates for international traffic between the United States and other countries (FCC 1996, 51). The FCC stresses the need for effective competition in countries negotiating accounting rates with US carriers, but it has proposed to establish a set of "benchmarks" for establishing these rates, and therefore settlement rates, in negotiations between US and foreign carriers. The FCC action has already begun to place downward pressure on these accounting rates, but rates remain far above the incremental cost of these international services.

Foreign Investments

Given the wave of privatizations, the remarkable changes in telecommunications technology, and the need for massive infrastructure investments to provide modern networks, many countries have opened their markets to foreign investment. In some countries, such as Jamaica, foreigners have been allowed to purchase the entire erstwhile nationalized company. In most countries, however, foreigners are not allowed to acquire a controlling interest in network companies, but rather are large minority investors in large national telecom companies or in smaller new ventures.

The US telephone companies are among the most aggressive of those investing in foreign operations. The Regional Bell Operating Companies (RBOCs), in particular, have major investments in New Zealand, Australia, Mexico, Chile, and Eastern Europe. (See table 6.2 for a partial listing. This listing reflects announced ventures from the trade press and other sources. Some ventures may have been abandoned. For a more complete discussion of the scope of these ventures, see Sidak [1995]).

Most of these investments appear to present few public-policy concerns. They certainly present no competition-policy issues as long as state authorities do not prevent others from entering to compete with them. Given the number of large potential investors in the world—at least 10 US companies, British Telecom, Cable & Wireless, NTT, France Telecom, Deutsche Telekom, and Bell Canada—it is difficult to see how any one company can exclude the others from similar competitive investments unless the host country decides to organize its own telecommunications sector as a cartel.

This is not to say that national governments might not cartelize their own markets in order to generate large capitalized market rents that bidders will be willing to pay to their treasuries, but such policies are bad per se, regardless of whether the buyers are foreign companies or domestic firms. It is not the foreign investments but the *domestic* communications policies that are often anticompetitive.

Table 6.2   Major foreign investments by traditional telephone companies

Participants	Type of business
Europe	
British Telecom/Viag (German utility)	Voice and data for corporate subscribers
Ameritech/Deutsche Bundespost/ Matav	Telephone service in Hungary
Ameritech/France Telecom/ Polish PTT	Telephone service in Poland
Bell South/Thyssen	Telecommunications services in Germany
Ameritech/Netcom GSM	Cellular service in Norway
US West/Olivetti Spa.	Regional cable networks in Italy
US West/EDS/France Telecom	Transactional and banking services
AT&T/Unisource	Uniworld
US West/Cable & Wireless	One 2 One—cellular service in the UK
US West/TCI	Cable service in the UK
US West/Deutsche Telekom/ France Telecom/Rostelkom	Telecommunications services in Russia
US West	Cellular service in Moscow
Sprint/Bulgarian government	Packet-switched network in Bulgaria
Air Touch/Mannesmann Mobilfunk	Cellular service in Germany
Sprint	Plessey Telnet
AT&T/Ukrainian Telephone Ministry/Deutsche Telekom/PTT Telecom (Netherlands)	Telecommunications service in Ukraine
Bell South/Thyssen/Vodafone/ Veba	Cellular service in Germany
British Telecom/Viag	Telecommunications services in Germany
SBC/Vodafone/CGE	Telecommunications in France
British Telecom/Banca Nacional del Lavoro	Albacom—Italian telecommunications services
Bell Atlantic/Air Touch/Mannesmann/ Olivetti	Cellular services in Italy

(continued)

Joint Ventures with Entertainment Media Firms

The third way in which the world's large telecommunications firms are entering joint ventures is through new service markets—in particular the markets for distributive and switched video. Telephone networks can be upgraded to deliver a large number of video signals by extending fiber optics or fiber optics and coaxial cables all the way to the subscriber's

Table 6.2 Major foreign investments by traditional telephone companies (continued)

Participants	Type of business
US West/Time Warner/ Multimedia Cable	Cable television in Spain
Air Touch/British Telecom	Cellular service in Spain
Air Touch/Telecel	Cellular service in Portugal
Western Hemisphere	
Cable & Wireless	Telephone service in Jamaica
MCI/Grupo Financiero Banamex Accival	Telephone service in Mexico
GTE/Grupo Financiero Bancomer	Telephone service in Mexico
SBC/France Telecom	Telmex—telephone service in Mexico
Bell Atlantic	Iusacell—cellular service in Mexico
Bell South	Cellular service in Guadalajara, Mexico
Sprint/Telmex	Long distance services in Mexico
AT&T/Grupo Industrial Alfa	Telephone service in Mexico
GTE/AT&T/two Argentine companies	CTI—cellular service in Argentina
Bell South	Cellular service in Venezuela
Bell South	Cidcom—cellular service in Chile
France Telecom	Telecom Argentina—telephone service in Argentina
Asia/Australia/New Zealand	
US West/Time Warner/ Toshiba/Itochu	Cable television in Japan
Air Touch/Cable & Wireless	Wireless service in Japan
Air Touch/TDP	Wireless service in Tokyo
NTT/Cable & Wireless	Personal communications services in Japan
Bell Atlantic/Ameritech	TNZ—telephone service in New Zealand
MCI/Bell Communications Enterprises	Clear communications—telephone service in New Zealand
Bell South	Cellular service in New Zealand
Telstra	Cellular service in New Zealand
Bell South/Cable & Wireless	Optus—telephone service in Australia
Vodafone	Cellular service in Australia

premises. But few telecommunications companies know how to market such services even if they can overcome the severe obstacles—the capital requirements and the regulatory opposition to such investments—to building such networks. As a result, the telecommunications companies have eagerly sought out joint ventures with motion picture companies, cable television companies, and major media firms, such as News Corp. (owned by Rupert Murdoch), Time Warner, or one of the major US television networks. The largest of these ventures are listed in table 6.3.

Table 6.3   Selected joint ventures in video and other media by large telecommunications firms

Participants	Type of venture
MCI/News Corp.	Investment
Nynex/Viacom	Investment
US West/Time Warner	Time Warner Enterprises (motion pictures and cable television)
AT&T/Silicon Graphics	Interactive digital solutions (interactive video)
Disney/Ameritech/Bell South/SWB	Video program development
PacTel/Nynex/Bell Atlantic	Joint video programming venture
AT&T/Paramount	Pilot programming for interactive video

Increasing Risks

The attempts by traditional telephone companies to diversify into new regional markets or new services involve substantial risks that have heretofore been alien to the telecom environment. These risks can be most readily seen in the capital-market and employment performance of US companies in the last 10 years. Similar changes could be demonstrated for the United Kingdom and, soon, for New Zealand, Australia, and Canada.

At one time, telephone companies in the United States were simply "utilities," or firms with an assured market, an assured return, a quiet life for the management, and steady but below-market returns for the widows and orphans who held their equities. Employment in telephone companies rose steadily until 1980 despite the fact that long distance services and terminal equipment had been liberalized in the 1970s and that AT&T was in court with the US government. The 1984 AT&T antitrust decree accelerated market liberalization and set off a series of stunning changes.

There is no better evidence of the changes now gripping telecommunications firms than the recent rise in the riskiness of their common equities. I have calculated the β coefficients from the capital-asset pricing model for the seven RBOCs, AT&T, and MCI over the past 11 years—the period since the AT&T divestiture. The results are reproduced in table 6.4.

For the first eight years of this period, the RBOCs' estimated β coefficients clustered very narrowly around 0.65, suggesting a systematic risk of about 35 percent less than that of the average US common equity. AT&T's β coefficient was somewhat higher and rose during the period, as was true of MCI's. Clearly, greater competition in the long distance market and riskier new investments, such as AT&T's purchase of NCR,

Table 6.4 The riskiness of US telecommunications equities since the AT&T divestiture (as measured by the capital asset pricing estimate of β coefficients)

Company	1984-89[a]	1990-91	1992-95
AT&T	0.74	0.83	1.01
MCI	0.87	1.24	1.47
GTE	0.71	0.77	0.81
Ameritech	0.58	0.56	1.03
Bell Atlantic	0.58	0.63	0.74
Bell South	0.75	0.55	0.68
Nynex	0.64	0.77	0.81
PacTel	0.67	0.43	1.16
SWB	0.66	0.60	0.66
US West	0.70	0.65	0.55

Note: These estimates are obtained from regressing the excess monthly returns from holding each company's common equity (defined as the dividend yield plus the rate of appreciation of the stock price less the 1-month yield on US Treasury bills) on the excess market return (defined as the monthly yield from holding a weighted portfolio of all traded equities less the US Treasury bill yield).

a. March 1984 through December 1989.

Source: The data are from Ibbotson Associates (1997) and the University of Chicago's Center for Research on Securities Prices.

have made these long distance carriers' equities riskier than those of traditional utilities. In the 1990s, however, the estimates of β coefficients have risen substantially for several of the RBOCs, particularly those that have been aggressively pursuing new strategies. Bell Atlantic, PacTel, and Nynex have been exploring and even making new investments in video distribution. Ameritech has begun to build cable networks in its region and has been the most aggressive in attempting to reenter long distance services by agreeing to open its local markets to competition. We now have the curious phenomenon that regulated "utilities" are often viewed as more risky than the average common equity in US markets (whose β coefficient equals 1.0).

In addition, the new competitive order in the United States has witnessed a shift of investment and employment away from telephone companies to new entrants and to customers' own networks. Since 1981, the decline in traditional telephone company employment has been 3.9 percent per year while all telephone industry employment has declined by about 1.7 percent per year. (These calculations are based on data reported by the FCC on employment by telephone companies—local carriers plus AT&T—and on the establishment data collected by the Bureau of Labor Statistics for the "telephone industry," Standard Industrial Classification (SIC) 481.) This suggests not only that technical progress and competition have shifted workers from the old-line companies to

Figure 6.1 Growth of net capital stock in the US telephone sector, 1970-1996

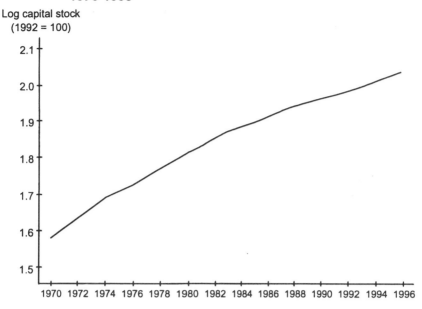

Log capital stock (1992 = 100)

new entrants, but that these changes have exerted substantial pressure on all telecommunications industry employment in the United States. The recent decision by AT&T to reduce its labor force by another 15 percent suggests that this process is far from complete, even in the United States.

Further evidence of the effect of market liberalization in the United States may be found in the growth of the capital stock in the "telephone" industry (figure 6.1). The growth in the telephone industry's capital stock has slowed since the mid-1980s despite the growth in local area networks, home terminal equipment, internet use, and so forth. The reason, in part, is that much of the growth is now in private business networks of equipment in the subscribers' premises. In addition, the Bureau of Economic Analysis may simply be missing much of the investment by new value-added carriers or business firms that are in a variety of related businesses in addition to telecommunications. If the United States has suffered employment reductions and a shift of employment and investment away from traditional telephone companies as it has liberalized its telecommunications sector, one must surmise that the effects of similar policies in other countries could be much more substantial. The postal service did not run AT&T; indeed, it was a private company, not a socialist enterprise. When liberalization occurs in Europe, for example, there could be much larger reductions in employment—reductions that could be blamed on the entry of foreign consortia that threaten domestic tranquillity and the security of the tele-communications network.

Antitrust Concerns

The replacement of national telephone monopolies, subject to government control, with liberalized markets in which firms wrestle for market share and competitive advantage has led to the substitution of competition policy for regulatory policy. The major concerns of antitrust authorities center around interconnection and abuse of local bottlenecks, but there is also lingering concern over backward integration by network firms into the provision of content.

Interconnection

Telecommunications is a network industry. For competition to occur, each network operator must be able to obtain access to customers and to terminate its traffic on networks serving other customers. If each subscriber has but one connection to the network, controlled by one network operator, this connection must be accessible to other operators if competition is to thrive. For this reason, most competition-policy issues involving telecommunications involve some concern over the abuse of a "bottleneck" position.

The expansion by government-owned or private regulated telephone monopolists into related markets often creates fears that these enterprises will use their monopoly bottlenecks to exclude competition in the new markets if the services require traditional telephone connections. The most obvious cases of such potential (or actual) abuses are those of traditional long distance services or the newer information and central-office services, such as voice mail, call forwarding, or automatic number identification. If local telephone companies offer all of these services, competition in these markets might be impossible unless regulatory or competition authorities have and enforce requirements for equal access to their networks.

Any liberalization of telephone markets, including those that involve the foreign investments and joint ventures described above, automatically draws attention to these issues of interconnection. National telecom authorities have had to deal with them in the United States, the United Kingdom, Canada, and Australia. New Zealand simply deregulated all telecommunications and invited the carriers to settle these issues through litigation.

It must be noted, however, that the monopoly power over local bottlenecks has nothing to do with the size of the players. Such power can be exercised by, say, the telecom company in Andorra against joint ventures involving AT&T, British Telecom, and News Corp., for example. These interconnection issues must be resolved in Illinois for prospective competition among Ameritech, MCI, and AT&T, just as they must be resolved for New Zealand for competition among consortia of large non-New Zealand and New Zealand players. One can only hope that most countries begin to understand the benefits of liberalizing interconnection for all carriers, including those wishing to offer international telecommunications.

Few countries appear willing to contemplate as drastic a step to control the exercise of monopoly power at the local bottleneck as that taken in the United States in 1984—namely, vertical divestiture of the major telephone company, AT&T. (For an analysis of the effects of the AT&T divestiture, see Crandall [1991].) The United Kingdom has expressly ruled out this vertical separation, but it has thus far prevented British Telecom from developing video services. Neither New Zealand nor Australia has embraced vertical separation, nor has Canada. However, the issue is very much alive in Japan where most observers expect NTT to be broken up into several companies.

The United States, on the other hand, has recently begun implementing a complicated new policy as the result of the Telecommunications Act of 1996. Under this statute, competitive entry into most communications markets is to be liberalized, but often with strict regulatory safeguards. Local telephone, long distance, and video-services markets, in particular, are to be liberalized. The local telephone companies, traditionally monopoly providers of network connections to all but the largest businesses, are required to "unbundle" their networks into a variety of network elements—local loops, switching, transmission, and signaling, among others—and to lease these elements to entrants at some measure of long-run incremental cost. (The precise measure of cost adopted by the FCC has been a matter of considerable dispute that has been turned over to the federal courts.) The local carriers must interconnect with entrants at any technically feasible point at rates that also need to be cost based. The RBOCs are to be allowed to offer long distance services between local access and transport areas for their own regions on a state-by-state basis once they have demonstrated that they have complied with the act's requirements for allowing entry into their local markets. Cable television companies, long distance companies, and any other organization may apply to become a certified local carrier. In addition, local telephone companies may now offer video services within their own regions. Unfortunately, the 1996 act creates a very complicated new regulatory regime that will require several years to implement because many of the act's more controversial aspects will inevitably be appealed to the courts. (For a more detailed critique of the 1996 act and its implementation, see Crandall [1997].)

Vertical Integration

The second most prominent antitrust issue is the integration of telecommunications companies with large media firms. All of the various communications technologies—telephone, cable television, terrestrial wireless, and satellite distribution—are being positioned to deliver a variety of one-way and switched-video services. As each of these distribution technologies develops the capacity to offer hundreds of video channels, the truly scarce technology may be the software to be delivered over these systems. Motion pictures, video games, interactive consumer services, and interactive medical services require large initial development costs. The winner in the race to deliver these technologies may be the firm that develops or otherwise obtains the rights to the most successful software in each of these areas. As a result, network owners have entered into numerous joint ventures or have even acquired outright many large software producers, as Turner and Time Warner have done.

The concern for competition authorities is that if one network owns a large share of the firms producing the software required for commercial success, potential new entrants will not be able to license this software on equal terms and will therefore not be able to compete. In the United States, this concern has been most evident in the relationship between cable television companies and television programmers (Waterman 1995). In the 1992 legislation that reintroduced cable television regulation, Congress inserted provisions requiring that all new distribution technologies be granted equal access to these vertically integrated programming ventures.

Although there is some limited empirical evidence of discrimination by a cable system in favor of programming in which it has an equity investment, the fear that existing network operators will gain control of program supply is often overstated. There is no "essential facility" in producing video programming. The talent required for its production can be augmented rather easily and cannot effectively be bound to long-term contracts in any case. Hollywood studios learned that decades ago when the Justice Department broke up their cartel in 1948. There are large numbers of film producers in the world. No television network, cable network, telephone system, or media mogul can possibly corner the market on new video entertainment.

Network companies may attempt to acquire control of other software producers as a bet on new applications for the switched-video information superhighway of the future. But no one can acquire control of Microsoft—other than Bill Gates—and attempts by network operators to corner the market on any new application would probably be short-lived. The ability of others to innovate around specific software solutions and the mobility and sheer numbers of those possessing the intellectual capital for such developments make monopoly control by downstream network owners very difficult.

The experience of two major Japanese electronics firms in acquiring large motion picture companies must serve as a warning to those who would venture into such territory. Sony's acquisition of Columbia Pictures and Matsushita's acquisition of MCA (Universal Pictures) cannot be judged as huge successes, much less threats to competition. Surely if, say, British Telecom or News Corp. would attempt to buy control of several large film distributors or even necessarily more modest-sized software producers, they could anticipate similar results. On a more recent note, US West's acquisition of a 25 percent interest in Time-Warner's motion picture and cable properties has surely not been a threat to competition in video distribution in the United States.

Although telecom mergers or joint ventures are not at the center of some of the recent giant media mergers, such as ABC/Disney, Viacom/Paramount, Turner/Time Warner, or CBS/Westinghouse, it is at least prudent to mention the possibility of increased media concentration in terms of technical developments in the telecommunications sector. It is at least conceivable that the dramatic changes taking place in microcomputers, fiber optics, and digital compression will lead to a single fiber carrying all communications to the final subscriber. The telephone company would be the likely owner of such a facility, perhaps leasing space to current cable companies or broadcasters as well as to independent information services. In such a world, the telecommunications companies could exercise control over the dissemination of ideas, political opinion, and other forms of expression. This control would exist regardless of whether the telecommunications giants acquired controlling interests in broadcasters, film studios, cable news networks, or other media.

The obvious antidote to such control of the essential channels of communication is the revival of common-carrier regulation with all of the economic inefficiencies that such regulation entails. But even to contemplate such a solution is premature given the uncertainties over the evolution of the relevant technologies.

National Control of the Communications Infrastructure

Other than the standard interconnection issues, foreign investments or joint ventures involving national telecommunications carriers raise few competition issues. However, the issues of national pride and national control often have a political appeal that can be exploited for the benefit of domestic carriers.

Virtually every country has some legal barrier to foreign investment in its communications sector. Many of these policies were developed in an earlier era when telecommunications was a monopoly, whether natural or otherwise. With communications technologies changing so rapidly today and with the integration of world markets for equipment, software, and telecommunications services, such laws are becoming a barrier to competition and even to the adoption of new technologies and services.

The cost of these barriers to foreign investment in communications infrastructure is most acute in smaller countries. The United States, with its 250 million people and $7 trillion economy, can easily draw on domestic resources to build competitive networks and develop new software and video programming. Yet the United States is relatively open to foreign investment, limiting the share of foreign equity participation (for no good reason) only if a company has licenses to use the electromagnetic spectrum. Canada, on the other hand, is only one-tenth the size of the United States, but it limits foreign investment in all communications ventures to 33 percent (GTE's ownership of BC Tel is an exception). At this moment, the Canadian government is anguishing over the inability of the original investors to succeed in developing Unitel, the second largest long distance company. AT&T, once a minority owner, has had effective control of the company thrust upon it, in apparent violation of Canadian law. Were the government to force AT&T to divest much of its interest, the company might well fail for lack of ability to attract Canadian capital.

An even smaller country, New Zealand, has allowed Bell Atlantic to become a major investor in its national telephone company and has permitted sizable foreign investment in the second company, Clear Communications. In addition, Bell South is a major investor in a cellular carrier in New Zealand. Australia has allowed similar foreign investment in its second carrier, Optus, and has allowed News Corp. and the national television networks to participate in new terrestrial cable networks being developed by Telstra and Optus. These smaller OECD countries appear to be cognizant of their need to allow foreign capital and expertise to assist in the very large task of building modern communications networks.

Given the risks inherent in developing new fiber/cable broadband networks or direct satellite broadcasting and particularly in developing the services to be offered on such systems, many governments will have to begin to allow a variety of joint ventures with the British, US, Japanese, German, Swedish, or Canadian firms that are in the forefront of developing such systems and services elsewhere in the world. Small countries will find it difficult to spread the risks of these new technologies strictly among ventures in their own markets.

The Road Ahead

It is very difficult to predict the future structure of telecommunications. Just five years ago, many pundits were predicting technological convergence of video, data, and voice over a single "information superhighway." This highway would have a fiber-optics backbone—a mixture of high-speed telephone switches and ATM switches—and coaxial cable drop lines to the final subscriber, delivering an all-digital signal to a subscriber's terminal, which would download and decompress video signals for television and deliver the remainder to other terminal devices, such as personal computers, telephones, and voice-storage machines.

Today, the convergence theory is being challenged. Telephone companies in the United States are slowing their deployment of hybrid fiber/coaxial cable systems. There is a real question as to whether current microcomputers can be equipped to provide the video jukebox services over these networks that were being envisioned just five years ago. Wireless telephony is rapidly gaining on terrestrial systems, and even orbiting satellites loom as new sources of competition. Perhaps most ominous is the development of three or four competitive direct-broadcast satellites, each capable of providing 200 or more channels of distributive-video services. Meanwhile, switched-video services are developing at a snail's pace—home banking, home shopping, and many other new services do not require massive bandwidth. Video games and remote CAT scans appear insufficient to support investments of $1,000 or more per subscriber for new switched-video systems.

All investments in current telecommunications networks and upstream software and programming are extremely risky. It is only natural that Time Warner, News Corp., Disney, AT&T, Cable & Wireless, British Telecom, the continental PTTs, the US RBOCs, the US commercial television networks, the large US cable companies, Singapore Telecom, China, Russia, Bulgaria, and the Ukraine are seeking international alliances to spread their risk and, in some cases, to attract needed capital. The sheer number of players, the rapidly changing technology, and the highly differentiated and changing services make it unlikely that the many contacts that these firms have with one another will have anticompetitive consequences. As markets are liberalized, however, the more important political threat is that formerly large national carriers will face substantial pressures to reduce employment, retire obsolete facilities, and even close many of their current operations. When and if full deregulation occurs, some of these firms may even face bankruptcy. One of the inevitable consequences of inviting full competition in a market once tightly controlled by government is that some firms cannot survive. The US airline and trucking industries offer abundant examples of such failures that are the price of allowing markets to function. Surely, erstwhile telephone monopolists, particularly those formerly operated by a national postal service, are in no less danger.

For the present, governments should build on the recent World Trade Organization agreement for telecommunications services by taking the following steps:

- Establishing clear rules for allowing interconnection of incumbents' and entrants' networks at prices approximating long-run incremental costs;

- Eliminating barriers to foreign investment in telecommunications network infrastructure;

- Privatizing national telecommunications carriers so that entrants need not fear that policymakers will unduly favor the incumbent carrier; and

- Allowing a variety of joint ventures among domestic and foreign carriers to enter the domestic telecommunications market.

References

Crandall, Robert W. 1991. *After the Breakup: U.S. Telecommunications in a More Competitive Era.* Washington: Brookings Institution.

Crandall, Robert W. 1997. *Are We Deregulating Telecommunications? Think Again.* Brookings Policy Brief 13. Washington: Brookings Institution (February).

Crandall, Robert W., and Leonard Waverman. 1996. *Cheap Talk: The Promise of Regulatory Reform in North American Telecommunications.* Washington: Brookings Institution.

Federal Communications Commission (FCC). 1996. Notice of Proposed Rulemaking. In *Matter of International Settlement Rates.* IB Docket 96-261 (19 December): 51.

Ibbotson Associates. 1997. *Stocks, Bonds, Bills, and Inflation.* Chicago.

Johnson, Leland L. 1989. *Competition, Pricing, and Regulatory Policy in the International Telephone Industry.* Santa Monica, CA: RAND (July).

Organization for Economic Cooperation and Development (OECD). 1990. *Performance Indicators for Public Telecommunications Operators.* Paris: Organization for Economic Cooperation and Development.

Organization for Economic Cooperation and Development (OECD). 1994. *International Telecommunications Tariffs: Charging Practices and Procedures.* Paris: Organization for Economic Cooperation and Development.

Sidak, J. Gregory. 1995. *Foreign Investment in Telecommunications.* Working Paper. Washington: American Enterprise Institute (November).

Waterman, David. 1995. Vertical Integration in Cable Television. Draft. Washington: American Enterprise Institute.

7

Global Networks, Electronic Trading, and the Rise of Digital Cash: Implications for Policymakers

JONATHAN D. ARONSON

The globalization of telecommunications and finance raises new competitive challenges for firms and regulatory challenges for policymakers. This chapter examines how the new global communications and computer networks are transforming the trading of foreign exchange and securities and the evolution of charge and credit card operations. It then suggests why and how these changes are critical to the future economic health of the United States and the wider global economy.

If the author's underlying thesis is correct—that the cross-section of telecommunications and finance will become one of the major drivers of economic growth—then policymakers need to pay careful attention. This does not mean that they should create a whole new regulatory infrastructure. Indeed, in industrial countries most regulators now concede that they cannot and should not try to micromanage economic developments. But despite the wishes of many in the private sector and the academic community, regulators are a persistent lot.[1] Regulators, therefore,

Jonathan Aronson is director of the School of International Relations and professor at the Annenberg School for Communications, University of Southern California. Technology outraced the evolution of this chapter. Inevitably, the data will rapidly be outdated, but the trends they illustrate may endure for some longer period. The author would like to thank Charles Schott, William Canis, and Kimberly Rupert for their help in arranging interviews and Claude Barfield, William Canis, Geza Feketekuty, Harry Freeman, Jim Larkin, Scott Loftesness, Michael Nugent, and Debora Spar for comments on earlier drafts.

1. This is one of the three iron laws that define business-government relations at the close of the century. The three are: Regulators may redefine their roles, but they never go away. Monopolists always welcome competition, but only in their competitors' markets. Business executives always claim that their sector is special and therefore deserves special treatment.

need to be smarter in what they try to accomplish and how they go about it. They need to understand when and how to intervene and when to stay out of the way and allow markets to function unimpeded. Although few specific policy prescriptions are presented here, the final section suggests issues, areas, and approaches that are likely to require more policymaking effort in the future, and in which policymakers should not become entangled.

The first section provides a brief overview of the historical links between information advances and the evolution of banking and finance. It suggests that the link between communications and finance has always been close but that in recent years new cross-sectoral competition has emerged. The second section examines how new global communications and computer networks are forcing transformations in foreign exchange and securities trading. The third section suggests that the changes in communications and computer networks are also revolutionizing the role and operation of the critical but little studied charge, credit, and smart card sectors.[2] The final section argues that the foreign exchange, securities, and card markets matter for national and international economic, monetary, and regulatory policies.

Telecommunications and the Transformation of Financial Markets

Advances in information and telecommunications technology affect business and foreign affairs (Hinckley 1989, 45-76; National Academy of Sciences 1988; Aronson 1991). Most bankers now accept that telecommunications holds together virtually all their technological capabilities. To illustrate, when Citibank established extensive private telecom networks, it gained significant advantage over rival banks.[3] In time, however, tardy competitors scrambled to catch up, and global networks went from being a competitive advantage to a commercial necessity (Office

2. Charge cards like American Express are paid off monthly. Visa, MasterCard, and other credit cards allow users to "buy now and pay later," but they charge interest on the outstanding balance. Smart cards contain an embedded microchip and require no instantaneous telecom interface with data banks stored elsewhere, but they require special merchant terminals to process them. Purchases using debit cards are deducted directly from checking accounts. Stored-value cards, like phone cards, replace cash; they are made in disposable and rechargeable versions. Secured credit cards require collateral to be deposited before purchases can be made.

3. Citibank and its former Chairman Walter Wriston were and continue to be at the forefront of efforts to integrate finance, telecom and information services (Wriston 1992; Interview with Walter Wriston, *Wired*, October 1996, 142-43, 200-5).

of Technology Assessment [OTA] 1992).[4] Indeed, when public network operators cut their prices to regain banking business, many banks shifted back to the public system (OTA 1992, 12).[5]

Timely information communicated at a low cost always helps bankers and investors. Financial integration based on steady improvements in communications technology has proceeded steadily from the late 17th century, when price quotes for securities and foreign exchange data for the London and Amsterdam markets were first published and exchanged (Neal 1990). With the installation of an underwater telegraph cable in 1851, market quotes and news began to pass freely between London and the continent. Four days after the first successful transatlantic cable went into operation on 27 July 1866, the *New York Evening Post* began publishing price quotations from the London exchange. By 1880, four years after the invention of the telephone, most brokers had telephones linked directly to trading floors, and stock tickers (invented in 1867) were installed in the offices of most New York banks and brokers (OTA 1990b, 129, 132).

More recently, international data connections have further advanced global banking and financial integration. By the mid-1980s, a "greenbelt" linked "the principal financial centers of the world—London, New York, Los Angeles, and Tokyo. Within it, a 24-hour-a-day, 365-day-a-year financial market operates with great freedom in financial transactions"(Cowhey and Aronson 1989, 5). Moreover, the World Trade Organization (WTO) Negotiations on Basic Telecommunications successfully concluded on 15 February 1997 will lower or remove domestic barriers to international competition in telecom services, push rates down, and help to integrate global financial and security markets further.

Today, new technologies are eroding the boundaries separating computer service firms, information network service providers, and data processors from the financial and communications realms. The line dividing content providers—particularly those involved with computer games, movies, and other forms of entertainment—from conduit builders and operators is also blurring. The rush is on for firms from previously different industries to compete. The possibilities are so intriguing that Seagrams, an old-line liquor company, sold its share of a proven money-

---

4. *The Economist* noted, "product technology will prove a great equalizer, as those banks that excel in it find it ever harder to keep ahead. In future, banks will need to offer their customers other ingredients as well if their products are to make money" (Capital Markets Survey, 21 July 1990, 15).

5. The telephone companies became alarmed when large customers (once treated simply as subscribers) began to set up their own telecom networks. Reluctantly, the operators offered virtual private networks to win back the business. Many customers returned to the telephone companies but still maintained a credible threat to bypass the public system if they did not receive favorable treatment.

maker, Dupont, to buy MCA, and an established name in the equipment industry, Westinghouse, has bet its future on the acquisition of CBS and its broadcasting network. Many of these efforts will fail or fall short of expectations. But the winners could transform the increasingly global, information-based economy.

Foreign Exchange and Securities Trading

International financial services became a major issue during NAFTA and Uruguay Round negotiations. The interest of world-class banking and insurance firms in open global markets was attracted by earlier efforts to create a unified European market and to liberalize the financial system in Japan. This interest carried over to regional and multilateral talks in the early 1990s. Meanwhile, the implications of gigantic flows of funds across borders, fueled by new technologies, drew the attention of concerned practitioners, policymakers, and scholars.[6] By 1995 the foreign exchange markets managed to clear something more than $1 trillion of trading a day.[7] The figure is now approaching $2 trillion a day. The huge amount of funds involved and the rapidity of their movement increased the risk financial institutions and traders faced and forced them to develop new risk management techniques. Some analysts now worry that fast money or the electronic economy could inadvertently destabilize the world's financial markets and create financial chaos. Britain's National Criminal Intelligence Service is concerned that near-perfect counterfeit copies of electronic money could trigger an international financial collapse. But offsetting these risks, the rewards presented by fast growing, technologically driven markets are immense.

Technological breakthroughs and widespread economic liberalization in banking also promoted greater global integration of transactions, payments systems, verification, settlements, record making, and other functions. They allowed banks and other financial institutions to improve payments capabilities and streamline processing operations. As banks

---

6. Joan Spero argued that advanced technology had helped financial institutions link markets and offer financial products globally, which contributed to the globalization of financial markets during the 1980s. (Financial deregulation and the volatility of prices, exchange rates, and interest rates also created a demand for innovative ways to hedge against, or take advantage of, new risks.) By 1988, Spero estimated, the value of money exchanged each week was equal to the annual volume of merchandise trade among nations (Spero 1988-89).

7. The Bank for International Settlements estimates that between 1986 and 1989, global currency trading doubled to $640 billion a day. The growth in currency trading has continued unabated since then (Randall Smith, How Currency Traders Play for High Stakes Against Central Banks, *Wall Street Journal*, 18 September 1992, 1).

and investment banks came to rely on information and communications technologies, the new systems became indispensable if financial entities were to stay competitive. Indeed, large investment banks now routinely spend hundreds of millions of dollars annually for their communications and information functions.

Three Major Changes

Three changes directly related to information and communications technology have altered the financial scene in recent years. The first is the rise of mathematical trading models developed and used by highly trained specialists. In an era of fixed exchange rates before 1971, foreign exchange trading was a sleepy backwater. Trading was generally sluggish and most currency crises were "telegraphed," giving traders ample time to move in or out of currencies in flux. The trading instincts of the traders were more important than formal education and models. The advent of flexible exchange rates after March 1973 made it possible for banks, firms, traders, and individual investors to make or lose huge sums of money in seconds. The speed and the stakes of the game increased. Suddenly, a firm's basic business could be overshadowed by foreign exchange shifts. The gentlemen traders who once dominated foreign exchange markets moved on, replaced by young, better-trained hot shots who rely on sophisticated mathematical models and may or may not have an intuitive grasp of the ebb and flow of markets (A Survey of the Frontiers of Finance: The Mathematics of Markets, *The Economist*, 9 October 1993).

A second major change that has revolutionized the financial community was the rise of global electronic networks that tied together markets and allowed round-the-clock trading of currencies, commodities, and securities. Networking of global markets proceeded in the policy vacuum that followed the collapse of the fixed exchange rate system. Early on, Reuters found that providing price information to foreign exchange traders was more profitable than its traditional news agency business.[8] In 1981, Reuters introduced new systems that allowed subscribers to execute transactions with other subscribers. Access to information from Reuters provided an important competitive edge. Its Monitor Dealing Service helped foreign exchange dealers negotiate transactions on terminals

---

8. Reuters takes price information from banks dealing in foreign exchange, repackages it, and sells it back to them for a fee. Telerate, owned by Dow Jones, was traditionally its major competitor and in the last decade Bloomberg LP (owned 30 percent by Merrill Lynch) has been increasing its market share among bond traders and by 1995 held 17 percent of the on-line business information market (Reuters Dives in All the Way, *Business Week*, 21 February 1994, 46-47; Brian Dumaine, Volume Down at Bloomberg, *Fortune*, 6 March 1995, 33-34).

instead of telephones. Within a decade the Monitor Dealing Service captured between 30 percent and 40 percent of the world market, executing hundreds of billions of dollars worth of transactions each business day. By 1990, Reuters had become the de facto global market for currencies.

By 1993, more than 50 percent of all foreign exchange transactions were negotiated directly by the two counterparties in the interbank market. Another 35 percent of the deals are matched by brokers, who bring together buyers and sellers for a price. Traders want to minimize this cost (Systems Prepare for Battle, *Financial Times*, 26 May 1993, Survey: Foreign Exchange, 3, 6). To meet this demand, Reuters launched an automated, screen-based trading system in May 1992, Dealing 2000-2, which allows traders to complete deals automatically and charge a much lower brokerage fee.[9] By 1996, about 40 percent of all spot currency trades were handled on 190,000 Reuters foreign-exchange terminals, making access to the Reuters network a necessity (A Smoother Ride, But Less Fun, *The Economist*, 24 February 1996, 77).

Predictably, large banks and phone brokers worry that Reuters will become too dominant in automated trading systems. So, encouraged by users, other firms launched competitive systems. In April 1993, a second system, Minex, a service heavily backed by Japanese banks and brokers as well as by Dow Jones Telerate, made its debut. (Dow Jones purchased Telerate in 1991 for $1.66 billion.) By mid-1994, Minex claimed to transact 6 percent of the daily dollar/yen broker turnover in Tokyo. In September 1993, 13 leading banks dealing in foreign exchange launched the Electronic Broking Service (EBS). EBS relies heavily on the captive business of its shareholders; they do not want to see a marked decline in the 60 percent of the foreign exchange business done by bank traders who call each other directly instead of going through a broker (Philip Gawith, Technology on the March, Special Section on Foreign Exchange, *Financial Times*, 2 June 1994, 4).

By late 1996, Reuters had almost 350,000 customers; Telerate, about 95,000; and Bloomberg, about 65,000. Momentum has swung to Bloomberg, the youngest of the three major firms, which provides almost real-time financial data and news, as well as comprehensive historical and analytic information about financial securities. Regardless of which firm triumphs, the volume of trading in traditional financial instruments will continue to soar. A wide range of new financial products will emerge as better communications and information technologies are introduced. Particularly in wholesale markets such as foreign exchange and derivative products, improved information networks now permit large-scale international

---

9. Dealing 2000-2 was designed to go beyond one-on-one negotiations between traders and emulate an auction market in which bids and offers by multiple parties are exposed simultaneously. The system displays the aggregate size of all bids and offers at each price, without disclosing the identities of participating dealers (OTA 1990a).

arbitrage and promote rapid integration between financial centers. Information technology is making it possible to reap large economies of scale but also promotes worldwide competition that continues to put pressure on spreads and profit margins. These two factors have induced financial service providers either to invest heavily in information technology or to contract for needed services (Hoekman and Sauve 1991, 124; Ken Auletta, The Bloomberg Threat: Why Does America's Newest Media Mogul Scare Dow Jones, *The New Yorker*, 10 March 1997, 38-51; Dow Jones Will Over Haul Telerate Unit, *New York Times*, 21 January 1997, C1, C6).

A third innovation revolutionizing the way financial institutions and individuals deal with money and trading is the cyberspace explosion represented by the rise of the internet and the world wide web. Electronic money and commerce on the internet are growing rapidly. With the arrival of increasingly secure internet payment systems, individuals and institutions have access to the same information and trading capabilities their brokers have. Soon, firms and individuals will regularly trade currencies in the spot and future markets without brokers. Proliferation and rapid development of new networks will expand and improve the ability of banks, other financial institutions, and their customers to exchange funds nationally and internationally. The evolution of these networks and markets raises new challenges for private parties and public regulators.

Global Networks and Securities Trading

Electronic trading networks now transcend national boundaries. Fifteen years ago, few US citizens or pension funds invested in overseas markets. There was little information available about foreign securities, and it was usually difficult and sometimes impossible to invest in them. Today, the trading of international securities is straightforward. The same information technologies that transformed foreign exchange and commodity markets now provide brokers and investors with real-time information and market quotations for most major global securities. It is fast becoming routine for brokers or individual investors to place their orders and instantly confirm their purchases and sales from their own computers hooked into the internet. Indeed, as long as the availability of funds is confirmed and appropriate encryption precautions are in place, it will soon be technically possible to trade in a security from any location anywhere in the world at any time of day or night. Thus an investor in Brussels can use a stockbroker in Sun Valley, Idaho, to buy securities in Hong Kong.

Global networks promoted internationalization of financial markets and spurred introduction of new trade offerings. Many major countries have liberalized their controls and lowered barriers to foreign investors. Even in countries where restrictions remain, country funds often allow foreign investors to participate in national economies. As financial markets opened, investors found it prudent to diversify their risks and opportunities by acquiring new types of securities (such as bundled mortgages or derivatives) and by internationalizing their portfolios (Capoglu 1990). By the early 1990s, investors worldwide traded about $1.1 trillion annually in securities outside of their home countries. In 1992, US investors purchased and sold foreign equities valued at $271 billion. Purchases have continued to rise, and new instruments such as derivatives are being traded (Beese 1994; The SEC's Open Door Policy, *Wall Street Journal*, 23 September 1993, 17; US Shares Could Suffer from Foreign-Stock Lure, *Wall Street Journal*, 23 August 1993, C1). Indeed, the expansion of the highly complex derivative market (which includes futures and options markets, swaps, and stock index options) was launched by computerization of trading and the information technology revolution.[10]

Electronic trading has been tried before. In the early 1980s, an equities trading experiment in London faltered, both because investors preferred to negotiate the price with a person over the phone and because market makers opposed innovations that would cut into their fees. Fifteen years later electronic trading is more attractive because cost is paramount and the wages of human traders continues to rise while the cost of using technology continues to fall. In addition, the ascent of program trading— in which the computers are more involved in the decisions to buy and sell stock—has given electronic trading a boost (Richard Waters, Reshaping the Trading Floor, *Financial Times*, 23 February 1993, 8; A Trading Floor on Every Screen, *Business Week*, 5 November 1990, 128-29).

In October 1986, the British government launched its "Big Bang," which allowed the action to move from the floor of the stock exchange to electronic trading floors. In one step, Britain deregulated the securities market, abolished fixed minimum commission rates, and allowed firms to operate as both traders and dealers (for the first time, they could trade both for themselves and their customers). British banks were allowed to underwrite securities and own brokerage houses. All restrictions on foreign ownership of British brokerage firms were lifted. Almost immediately huge sums of new money flowed into the London Stock Exchange, and a wave of mergers and acquisitions rewrote the financial landscape (OTA 1990a, 43-44).[11] Within four years "turnover in

---

10. The derivatives market still is not well understood by ordinary investors. Large losses that accompanied the rise in interest rates in the first quarter of 1994 brought wider attention to this market and raised questions about whether and how regulation of this global market should be restructured. This market relies heavily on large institutional investors seeking to tailor their portfolios of assets to advance their interests and diversify their risks.

11. By the same token if the US Congress approves new legislation that remove barriers separating banking and securities firms, the United States could experiences its own "Little Bang" (Fuse Lit for Little Bang, *Financial Times*, 3 May 1995, 13).

London of foreign equities exceeded that of domestic stock for the first time," primarily a result of the growth of the London Stock Exchange Automated Quotation International system (SEAQ International)—the London Stock Exchange's screen-based market for trading foreign equities. Today, the London Stock Exchange is probably the most global stock market in the world.[12]

One successful attempt to exploit these changes was Reuters' Instinet, an alternative electronic broker-dealer offering an electronic securities trading system for doing business on existing exchanges. Instinet competes with other brokers but does not trade for its own account. By using Instinet rather than existing brokers, customers have full access to all the buy and sell orders Instinet has received from customers dealing on the exchange. By contrast, Tradepoint, a British exchange that opened for business in September 1995, tries to compete against the London Stock Exchange. Tradepoint transacts trades electronically, with computers bringing together buyers and sellers. Tradepoint is "an order-driven system: anyone who wants to buy or sell shares posts an order on a screen, then waits to see if anyone comes forward to trade against the order electronically" (O'Brien 1992; Shaping Up, *The Economist*, 23 March 1996, 75).

On the other side of the globe, the slowness of Japan to internationalize the yen and open its foreign exchange and securities markets provided a tremendous boost to efforts by Hong Kong and Singapore to establish regional financial centers. In the mid-1980s, under pressure both from Japanese business and foreign firms and governments, Japan's Ministry of Finance began liberalizing. Only then did Japan rapidly move to the forefront of Asian financial markets.[13] Although the Tokyo Stock Exchange rivals the New York Stock Exchange in volume, it is not nearly as internationalized as London or New York. The Tokyo Stock Exchange listed its first six foreign stocks in December 1973 and reached 127 stocks in 1991 before retreating to 82 in early 1995, reflecting the high cost of maintaining a listing and the falling number of Japanese-based shareholders interested in non-Japanese stocks (Tokyo Exchange Hit by Departures of US Companies, *Financial Times*, 26-27 March 1995, 1). Electronic advances have also pushed the virtual integration of stock markets. Market integration was promoted by the desire to reduce transaction costs, manage currency risks, and make certain that equities quoted on several exchanges

---

12. In Amsterdam and Frankfurt fewer foreign equities are listed than in London, but the percentage of foreign equities traded is higher (O'Brien 1992).

13. Significant financial problems in Tokyo made the market less attractive. In addition, there remain significant language and cultural barriers that hamper foreigners wishing to operate in Tokyo and the cost of setting up and recruiting top-rank Japanese staff is considerable (US Brokers' Stock Is Rising on the Tokyo Exchange, *Business Week*, 9 July 1990, 60).

are quoted at the same price in different places at the same instant. Fearful that supranational authorities might enter the field, stock markets pushed ahead to unite. In many countries, regional markets are merging into national exchanges, and national exchanges are consolidating.

Integration is underway in Europe. The European Commission, in an effort to create a unified European securities market, issued two directives. When fully implemented, the investment services directive and the capital adequacy directive are expected to foster a more integrated European capital market.[14] Eurolist, a project of the Federation of European Stock Exchanges, expects to provide a one-stop-shopping facility. Large European firms listed on several exchanges will use standardized procedures and provide Europe-wide dissemination of company news. Competitors on the horizon may push integration further. A new network that would allow after-hours transaction by institutional investors is being developed. The four Scandinavian exchanges will offer Nordquote, which will allow intermediaries to indicate bid and ask quotations in about 100 Nordic stocks over the phone. In addition, Instinet, the Reuters order-driven system, and Tradepoint are likely to push the exchanges to explore greater integration (Distler 1994, 21-26).[15]

Two Challenges

These developments raise at least two major challenges. First, the technology will change the role of traders, stockbrokers, and other middlemen. The need for stockbrokers to place buy and sell orders and maintain records for investors will decline. The transactions themselves will be subject to competition, driving down fees for executing trades. Already, discount brokers allow investors to play the market more aggressively because they can afford to move in and out of stocks swiftly. For the sophisticated, or foolish, investor, who tracks and trades stocks and other equities for himself electronically, the broker becomes unnecessary. Although new technologies alter the role of the brokers, the occupation is unlikely to disappear. Stockbrokers will become investment advisers and money managers. Because most people have neither the time nor the expertise to trade extensively, they will still rely on professionals to advise and trade for them. Indeed, novice investors trading electronically in equities around the world will be gambling, just as if they were playing slot machines. Rather than gamble, many investors will want others to

---

14. Implementation, however, has been spotty (Even Now, They Shall Not Pass, *The Economist*, 16 December 1995, 71-72).

15. In January 1996, the Paris Bourse announced that it would soon trade shares in British, Dutch, Italian, Spanish, Swedish, and Swiss companies. It hopes that shares in Europe's 500 largest firms will eventually be traded in Paris (*The Economist*, 23 March 1996, 75).

advise and trade for them.[16] But the value stock brokers will need to provide is expertise about firms and markets, not the ability to trade securities efficiently.

Second, the new technology raises important questions for national regulators. If traders can trade in any security, commodity, or currency from any location, this raises new international regulatory issues related to prudence, probity, and oversight. National regulators likely will need to cooperate to harmonize their regulations that require disclosure by companies. Otherwise, US regulators may exercise an extraterritorial reach to protect their investors from fraud in overseas markets. The definition of a level playing field in an age of integrated securities markets will evolve rapidly in the next few years. These issues are considered in the concluding section.

Charge, Credit, and Smart Cards

Technology also is transforming national and international commerce. In the United States today, credit and charge cards and automatic teller machines are ubiquitous. By 1995, fewer than 1 percent of bank customers managed their money on-line, but telephones were used for a quarter of personal banking transactions (Surf's Up for New-Wave Bankers, *The Economist*, 7 October 1995, 77; Timothy L. O'Brien, On-Line Banking Has Bankers Fretting PCS May Replace Branches, *Wall Street Journal*, 25 October 1995, A13). It is likely that as the internet and world wide web continue to attract users, they will change the banking and financial scene. The volume of funds involved in the card sector is significant and growing. In 1995 "consumers spent $1.6 trillion with plastic, almost twice as much as in 1992" (A Punch-Up in Plastic, *The Economist*, 8 June 1996, 77). Improved communication infrastructure is making the charge and credit card business and the emerging debit and smart card industries more important for operation of financial systems and thus for national regulators.

Competition in the Charge and Credit Card Industry

The leading card issuers—Visa, MasterCard, and American Express—compete fiercely but are quite different in the basic products they offer. Visa and MasterCard are bank-based systems. Their credit cards are

---

16. The danger is analogous to what followed the rise of the self-service gas station-convenience store and the disappearance of most full-service stations. The consumer finds it easy to pump his own gas, but often neglects to check his oil, water, and tires. Large repair bills frequently follow.

issued to customers by more than 20,000 banks worldwide, so there is no direct relationship between the customer and Visa or MasterCard. Unless the issuing bank has a strong international presence, such as Citibank or Bank of America, customers do not have access to a worldwide office system.[17] In addition, customers are assigned a specific credit limit, and as much as 90 percent of the income of Visa and MasterCard is earned from interest on the credit balances of their customers. If a customer's balance exceeds his or her limit, the card is refused. Indeed, since banks may charge customers high interest rates on unpaid balances, banks prefer customers not to pay off the principal so long as they make regular payments to cover the interest.[18] By 1984, 71 percent of all Americans between 17 and 65 carried a bank card of some kind.[19] By 1995, Visa and MasterCard controlled 74 percent of credit card purchases in the United States, up from 69 percent in 1990. The share American Express controlled fell from 24 percent in 1990 to 18 percent in 1995. About half of its charge volume came from outside the United States, where its share of the global market fell from 13 percent in 1992 to 10.4 percent in 1995 (Laurie Hays, American Express Sizes Up Rivals, Turns Green, *Wall Street Journal*, 3 May 1996, B1; A Punch-Up in Plastic, *The Economist*, 8 June 1996, 77-78). To put this in perspective, in 1994 the consumer payment sales volume of Visa International reached almost $631 billion, up 22 percent from $519 billion in 1993.[20]

---

17. This drawback has faded as brick-and-mortar offices grow less critical and telecommunications is used for all manner of customer services except cash. For cash access, the Plus-Cirrus ATM networks provide Visa and MasterCard holders more cash outlets than American Express.

18. Consider the customer who charges an average of $2,000 a month on a Citibank Visa card linked to American Airlines and pays off her full balance on time each month. During the course of a year the customer gains on the float from delayed billing and earns two free round-trip tickets within the United States on American Airlines. Citibank provides the customer with a float but also must pay American Airlines about 1 percent of charges to compensate it for issuing frequent flyer mileage. Citibank does, however, collect an annual fee from the customer, fees from merchants, and derives valuable information about the spending habits of its customer, which may be resold to others.

19. In the early 1990s, propelled by "cobranding" of cards with huge nonfinancial companies such as AT&T and Ameritech, General Motors and Ford, American Airlines and United Airlines, and the AFL-CIO and the American Association of Retired People, MasterCard began to close the gap with Visa (Kelley Holland, Visa Volleys for Market Share, *Business Week*, 27 September 1993, 46; The War of the Plastic, *Business Week*, 15 April 1991, 28-29; Melting Point in the Plastic War, *Fortune*, 20 May 1991, 71-75).

20. The number of Visa cards worldwide reached 391 million in 1994. The 1994 figures were distributed geographically as follows: United States—$291 billion from 206.2 million cards; Canada—$33 billion from 18.2 million cards; Europe, Middle East, and Africa—$212 billion from 77.9 million cards; Asia-Pacific—$73 billion from 69 million cards; and Latin America—$22 billion from 19.7 million cards. Note, Americans tend to hold more cards per person than consumers in other regions, but consumers from Europe, the Middle

American Express, however, was built on travel and entertainment charge cards, not mass-market credit cards. Because of its narrower focus, the growth of American Express lagged behind both Visa and MasterCard. It has a closed-loop relationship with the establishments that accept the card and with charge card members. Card members have no set credit limit, but American Express reassesses credit authorizations based on the member's past purchases and credit rating. Card members are expected to pay off their entire monthly balance each billing period. Thus American Express makes the bulk of its fees from merchants, which in turn generally pay higher fees on American Express card sales than on sales charged to competing credit cards. In early 1987, American Express entered the credit card market with its Optima card. To try to win customers away from the bank credit cards, it charged users a lower interest rate than most banks. The total number of American Express cards in use actually declined from a high of close to 37 million in mid-1991 to an average of 35.4 million cards in 1993. Less than a third of these cards were held outside the United States, and only $34 billion out of the total American Express charge volume of $124 billion in 1993 originated outside the United States.

ATMs, Information Networks, and the Card Industry

Advances in telecommunications and information networks have helped the card business grow by about 20 percent annually since 1980. The availability of low-cost, point-of-service terminals, which use switched telecommunications services to allow rapid authorization of purchases, make cards simpler than checks and almost as simple as cash. This ease of use and the timely processing of card transactions are keys to success in the card business. Technology also altered the competitive terrain between American Express, the bank-based systems, and more recent ventures such as the Discover Card and the AT&T Universal Card, which promised discounts on phone calls and no annual fee to early subscribers.

An earlier example of technology's effect on the card industry is illustrative. The introduction of automatic teller machines (ATMs) transformed the card industry. The first through-the-wall cash dispenser began operation near London in 1967. Two years later, automatic teller machines appeared in the United States—a dividend of technology for electronic funds transfer. Beginning with American Express Gold Cards, credit card customers could use their cards to get cash from American Express cash

---

East, and Africa charge on average almost twice as much as their American counterparts per card (Visa International Press Release, 5 April 1995). I thank Scott Loftesness for bringing this announcement to my attention: "In 1995 American Express's card volume grew by 15 percent and Visa's volume by 25 percent" (The Battle Over the Credit Cards, *Forbes*, 1 July 1996).

dispensers and have their bank accounts instantly debited. But because there are far more banks than American Express offices or terminals, a huge opening was created for Visa and MasterCard. They began working through ATM networks such as Cirrus and Bank Plus to make cash easily available to their own cardholders. Thus, even though American Express pioneered the field, ATMs quickly put American Express on the defensive. Through sharing agreements, American Express responded by providing its cardholders access to more than 100,000 ATMs worldwide. Soon, smarter ATMs—"all-purpose multimedia kiosks"—probably will replace pay telephones and make it possible for banks, telecommunications firms, and other entities to provide customers with telephone, fax, computer, and financial links paid for with smart cards (Max Glaskin, Smart Card Meets Ingenious Terminal, Survey of Computers in Finance, *Financial Times*, 15 November 1994, 6).

Today, the key to continued fast growth and globalization of the card business is also the ongoing construction, expansion, and modernization of the communication and information infrastructure that backstops everything. Use of a plastic card initiates a transaction. All the rest of the support is delivered through technology, including movement of information and management of transactions. The card companies all maintain gigantic private networks that link every major metropolitan area. Collectively, they spend something on the order of $500 million each year on technology so that they can quickly and efficiently process millions of transactions each day. The banks associated with Visa and MasterCard spend at least that much to update their own card technology.

At the heart of these systems are consistent architectures that within seconds can verify that a card is valid and that the user has sufficient credit or funds to cover the transaction. This authorization allows merchants worldwide to accept a card as payment because by obtaining an authorization, the merchant obtains a guarantee of payment from the bank card system. Parallel networks to clear funds instantaneously are less developed because, until more debit cards are in circulation, funds will be cleared at intervals through credit card payments. Payments to the banks, the merchants, and American Express are made by mail, bank deposits, and electronic transfers.[21] Besides moving information and funds around with impressive speed and precision, these networks also generate huge amounts of customized information about the buying habits of individual customers. At first, network developers excluded their competitors, but because significant overcapacity emerged, most of them were opened to generate additional revenues. Cooperation between Visa, MasterCard, and American Express also was pushed by the need among

---

21. The clearing-settlement system often shifts the funds from the card issuer to the merchant's bank to accomplish the payment to the merchant.

merchants for common technologies and by the need among card issuers to control costs, improve security, and avoid redundancy.

Technology can also be used to discourage competition and trade. For example, banks in Canada, France, and elsewhere did not allow American Express card holders to use bank ATMs to withdraw cash.[22] And in the mid-1980s, the French government initiated a program to wipe out duplicate payment systems so that a single electronic railroad would be universally available. The French government promised to make electronic funds available to all, but the French banks quickly joined together to form Groupement des Cartes Bancaires (Bank Card Group), which worked to bar access to cards they considered competitive with their own. Thus, French bank ATMs will accept foreign-issued American Express cards but not those issued to French residents.[23] Not surprisingly, American Express views such impediments as unfair trade barriers and lobbies the US government to intervene on its behalf.

Debit Cards, Smart Cards, Internet, and the Future

Not all new technologies are quickly embraced. In the United States consumers have resisted debit and point-of-sale systems. Personal Identification Numbers (PINs) on credit cards and smart cards with embedded microchips are commonplace in Japan, France, Germany, and elsewhere in Europe. Since the 1970s more than 250 million smart cards were made in Europe, mostly prepaid disposable telephone cards. Analysts estimate that in 1997, 90 percent of smart card use was concentrated in Europe and only 2 percent in the Americas, but they expect smart cards to take off in the United States and throughout the world. The United States could account for 20 percent of smart card use by 2001. Thus the dream of the cashless society, like the vision of a full merger of communications and computer technology, has been slower to dawn than boosters

---

22. Except for the Bank of Montreal, which had a sharing agreement with American Express, all the large Canadian bank ATM networks excluded American Express. Canada also wanted to charge American Express $7.50 per card to clear transactions, a fee American Express considered excessive and discriminatory. Until recently, the Canadian government required that all records on Canadian customers remain in Canada. Records now may be moved out of Canada as long as a backup copy remains in the country.

23. When American Express and Credit Lyonnaise reached an agreement that would have granted American Express card holders access to the Credit Lyonnaise ATM network, the Groupement intervened. It added a stipulation that only American Express card members with cards issued outside of France could use the Credit Lyonnaise system. Technological barriers, therefore, made it more difficult for American Express to develop a French client base. Until late 1994, American Express also faced a problem in Saudi Arabia, where the government and the local banks worked together with Visa and MasterCard but made it difficult for American Express to develop a foothold. This information is based on interviews by the author.

first predicted. Although it is often practical to debit the customer's account directly for purchases, most consumers resist giving up the free float and the power to challenge shoddy merchandise or services by withholding payment (Why Warren Buffett's Betting Big on American Express, *Fortune*, 30 October 1995, 74; The Ultimate Plastic, *Business Week*, 19 May 1997, 119-22).

An interim variant of the debit card might be disposable cards denominated in flexible values, similar to phone cards and subway cards. When a customer goes to the bank or other terminal linked to her bank account, she might ask for a disposable card instead of cash. The card would be smart enough to know how much value it held and to pay and receive money when it was used at the right terminals. A PIN number could be automatically encoded onto the card to prevent others from using it. There is no reason that such a card could not generate records of its use for the holder. Such a system, dubbed Mondex, started testing in July 1995. Its two British founders, National Westminster and Midland Bank, hope that its technology will be accepted as a global standard for digital cash. They predict that Mondex could eventually capture 40 percent of the total cash market by sparing banks and merchants the laborious task of counting and recounting money. CAFE, a European Commission-sponsored project, also hopes to offer money on smart cards.

By simplifying electronic delivery of cash, smart card issuers may shuffle competition in the banking industry, allowing some banks to lower their costs dramatically and compete with established branch franchises (The Smart Card Cashes In, *The Economist*, 29 January 1994, 73-74). As this type of card becomes more common, it might also reduce street crime. If people generally carried debit cards with PINs instead of cash, perhaps they would be somewhat less attractive victims. Such cards also would be highly attractive in places like Russia, where checks do not exist and "until now the only way to pay for things has been to carry around huge amounts of devaluing rubles" (The People's Plastic, *The Economist*, 12 February 1994, 83).

Simplified, universal debit and smart cards are beginning to make headway. Although there is some evidence that ATMs have increased the convenience of cash more than credit and debit cards have made inroads, it is likely there will be movement toward a cashless and even a checkless society. Earnings will be safely and automatically deposited electronically. Automatic payment systems will pay phone, utility, and other bills as they come due. At the Olympics in Atlanta in July 1996, Visa tested a new cash card with a chip embedded in the card, Visa Cash, that substitutes for cash. As a consumer uses the card, its value goes down. No PIN or password is required. The next stage is likely to be introduction of reloadable cash cards (James Gleick, Cash Is Dying, *The New York Times Magazine*, 16 June 1996, 29). Eventually, people probably will carry smart cards with individual personal identification numbers (or their future fraud-proof equivalent) to make misuse of cards more difficult.[24] Smart cards seem likely to come in three varieties: an electronic purse—essentially a traveler's check that makes exact change and is discarded when it is debited down to zero; a rechargeable card; and a personalized card that holds a computer chip with a digital picture, DNA signature, or other form of identification. Microsoft's Bill Gates, for example, envisions a "wallet PC" that would hold, among other things, electronic versions of credit cards and money. Secure, coded messages transmitted between the user's wallet PC and bank would allow purchases without money or credit cards (Interview with Walter Wriston, *Wired*, October 1996, 142; Bill Gates's Vision, *Business Week*, 27 June 1994, 60). Customers already are using credit and debit cards extensively in supermarkets. In the future, they are likely to use debit cards or electronic money to buy everything from morning coffee to a new car.[25] The user's bank accounts will be debited automatically, and detailed purchasing breakdowns can be generated for the customer (to facilitate budgeting and tax preparation), the government (to monitor tax returns), and interested third parties (concerned about credit worthiness or marketing new products). Not all consumers, of course, will embrace these innovations and the associated loss of privacy, but the transactional benefits probably will eventually win substantial support.[26]

In addition to smart cards, many believe that there will be a booming market for electronic money (or E-cash) that will be exchanged over the internet. The first reasonably secure internet payments system, First

---

24. France is the leader in smart cards, and the French are far more accepting of debit cards. Competition made progress slower in the United States, but over time the United States will probably catch up.

25. The branding of credit cards and other ventures is in the works. "Green stamp" bonuses are used to entice buyers to use them instead of cash or checks. For example, customers now receive frequent flyer miles when they buy groceries or charge telephone calls. One striking example occurred when business executive and art collector Eli Broad purchased a painting by Roy Lichtenstein at auction for $2.5 million. By using his American Express card he earned 2.5 million frequent flyer miles on Delta Air Lines at the same time.

26. Innovations are often slow to be adopted. For example, customers at first spurned bank ATMs as impersonal; now they prefer the convenience and efficiency. Many people decry the loss of personal privacy made possible by data banks and resent the constant bombardment of mail and phone solicitations. There is greater concern that information collected by credit bureaus can be so easily accessed. There also exists the specter of companies that offer lawyers, insurance companies, prospective employers, private investigators, and others the ability to rapidly search public information bases on everything from court records to bankruptcies to mortgage payments for a minimal fee. In the past such public information was available but hard to sift. Although search companies try to screen scam artists, people will increasingly live their lives in fishbowls. Proper discretion may limit the extent of abuse.

Virtual Holdings, began operation in October 1994. Open Market Inc. and Wells Fargo Bank-Cybercash formed an alliance with Virtual Vineyards to sell wine and started commercial testing in February 1995. Visa has teamed with Microsoft, MasterCard has partnered with Netscape, and financial institutions such as BankAmerica, KeyCorp, and NorWest have announced that they, too, will offer similar systems. In May 1996, Deutsche Bank launched a program with internet pioneer DigiCash to test the use of E-cash. The trial involves loading digital cash into computers and simultaneously debiting the individual's account (Deutsche Bank-DigiCash Plan "E-Cash" Trial, *Wall Street Journal*, 10 May 1996, A5).

DigiCash uses public key encryption to send sensitive information over the internet. DigiCash features special blinding technology that lets banks certify an electronic note without knowing to whom it was issued. In other words, the bank knows the payer (its own customer whose account is debited) but not the payee. Unlike an encrypted credit card transaction, the electronic cash is as "anonymous as paper money" (The Future of Money, *Business Week*, 12 June 1995, 69). Even more dramatic is Citicorp's Electronic Monetary System, an infrastructure for using electronic money issued by Citibank and other banks. Market researcher Killen & Associates boldly predicts that $300 billion worth of goods and services will be sold annually over the internet by 2000 (Safe Passage in Cyberspace, *Business Week*, 20 March 1995, 33; The Future of Money, *Business Week*, 12 June 1995, 66-74). Unlike the smart card, which is a product seeking a market, the internet is a vast market of tens of millions of internet users and tens of thousands firms that is growing 10 percent a month and actively seeking real products. It is not yet clear whether electronic money offered over the internet will succeed or suffer the so far inglorious fate of home-banking experiments. If it succeeds, electronic money is likely to have significant implications for money, commerce, and regulation in the non-electronic world (Electronic Money, *The Economist*, 26 November 1994, 21-23). Banks and regulators will face daunting new challenges related to issues of privacy, security, tax evasion, money laundering, counterfeiting, and other financial crimes likely to be committed on the internet and other private networks. Indeed, when then Federal Reserve Vice Chairman Alan Blinder was asked after a March 1995 speech whether or not the Federal Reserve was studying the regulatory issues surrounding digital cash, he responded, "Digital what?" After a moment he added, "It's literally at the thinking stage" (The Future of Money, *Business Week*, 12 June 1995, 78). A task force chaired by the comptroller of the currency is examining these issues but has not made any serious recommendations.

In short, the dramatic adjustments introduced by improved communications and information technologies in areas such as foreign exchange and commodity trading, securities trading, and the card industry spanned the globe with impressive speed. These changes raise new and important international monetary and financial issues not previously much considered by policymakers, political economists, and economists concerned with international monetary and financial developments.

Implications for Policymakers

So far this chapter has explored how new telecommunications and information technologies nurtured the evolution and operation of the foreign exchange, securities, and card markets. This section explores what these developments mean for policymakers and regulators. Where money and communications intersect, thorny challenges arise for different kinds of government policies at national, regional, and global levels. Some believe that control of the money supply is just the tip of the iceberg. The mobility, volume, and electronic nature of international financial flows could undermine the autonomy and effectiveness of national macroeconomics and monetary policies. Governments may also need to rethink their trade, regulatory, and antitrust policies to adjust to new developments.

It will be necessary for states to revamp and reform their national policies. This section starts by questioning whether national policies and regulations can still be effective and efficient and examines challenges at the national level that arise because the major capital markets are becoming electronically and globally interdependent. One solution would be for policymakers and regulators to cooperate more closely across national borders. Under some circumstances policymakers may try to work cooperatively to craft regional or global policies. To carry out such policies, improved coordination of macroeconomic policies among industrial countries may be needed. But calling for cooperation and coordination is a far cry from achieving it. Moreover, global cooperation and harmonization of policies and regulation is not always effective or desirable. Sometimes governments should liberalize their regulatory control and allow competition within markets to ensure that consumers are well served. If governments cannot competently monitor and regulate markets, perhaps they should stop trying and cede at least some control to the marketplace.

National Policymaking Challenges

To demonstrate how the new relationship between finance and telecommunications may raise serious challenges for national policymakers, it is best to examine three issues: efficiency of economic and regulatory policies, management of payment risks, and protection of privacy.

Global markets raise twin challenges, as they do for the efficiency of national economic and regulatory policies. First, policies and regulations conceived for national markets inevitably are less efficient when applied globally. Second, international developments intrude more and more on national economies, making it increasingly difficult for officials to regulate their commerce and fine tune their monetary policy. It is sensible, for instance, for financial institutions to use overseas units and international markets to legally circumvent national requirements. Companies also can shift vast sums of money to minimize their tax responsibilities by earning income and holding their assets beyond the tax collector's grasp. Just as the passage of credit controls in the United States in the 1960s propelled the growth of the Euromarkets, national attempts to regulate banks, stock markets, and card operations sometimes can prove counterproductive. Moreover, cross-sectoral regulatory issues that bridge telecommunications and finance become increasingly important when these sectors are liberalized at different rates. When banks and telecommunications firms trespass into each other's realms, often it is difficult to decide who should regulate. Thus, when applied to networks, policies that promote competition and discourage collusion often become confused. It is unclear, for example, whether shared networks reduce competition among banks or whether network suppliers compete directly with banks. Thus, the Office of Technology Assessment (OTA) notes that if "payment systems are viewed as telecommunications networks rather than as banking networks, any third party can provide switches to route money transfers from one location to another across national boundaries" (OTA 1992, 28). In short, as boundaries separating industries blur, in part because of the impact of new telecommunications and information technologies, national policies and regulation and will be more difficult to design, implement, and enforce.

To manage payment risk, banks routinely use technology to credit and debit their own accounts (and the accounts of their customers) without physically transferring currency. Unless electronic trading and information networks are fast, reliable, and secure, the participants will worry about payment risks. If settlements are flawed, banks will not receive their funds quickly, undermining confidence in electronic trading. Some argue that, to guarantee the timely clearing of transactions, it is important to convert promptly to paperless settlement (Group of Thirty 1989). However, although the rise of automated trading and universal credit and debit cards create the potential for an improved audit trail of transactions, nobody has yet figured out how to create a surveillance system to monitor the vast global trading system. And as the stream of money swells, the temptation for larceny grows, the potential for mishaps rises, and the danger soars that payment losses could spark a major bank failure that cascades through the economy. When German bank supervisors closed Bankhaus Herstatt, a single small bank in Cologne with open spot transactions, in 1974, the system shook. That crisis was small compared to the problems now possible.

As banks rely on international telecommunications to transfer funds, payment risks grow. By mid-1992 CHIPS, the bank-run, New York-based clearing system, handled more than $900 billion in transfers each day. Total overdrafts at the peak of the business day averaged $45 billion. Unlike the Fedwire electronic payments system the Federal Reserve System operates and backs, if one of the CHIPS banks failed or could not meet its obligations at the end of the day, the other banks in the system would be forced to cover the defaulting bank's debit position.

Central banks also want to make certain there are no settlement failures of banks using the Brussels-based SWIFT system. They want to prevent payments problems from cycling from bank to bank, causing systemic collapse. The Bank for International Settlements has warned that of all the types of risks "to which banks may become exposed through the accelerated use of the new technology, it is this systemic risk that is the greatest cause for concern" (BIS 1989, 34).

National regulators are working together to plan ways to minimize such risks and control problems that do arise. However, as corporations begin to use their Electronic Data Interchange networks to transfer money back and forth across borders and between institutions, worries about payments risks may increase, and the challenge to minimize such risks will loom large. Regulators do not want to curtail market growth, but they are determined to minimize the risk of future crises.

The problem of international payment risk is compounded when criminal activities are considered. Bank and stock market fraud, credit card villainy, and electronic commerce larceny are growing problems for financial institutions, card issuers, and the internet. These threats have prompted ever more elaborate efforts to ensure account security and have sparked interest in more serious prudential regulation in the United States, Canada, Japan, France, and Germany. Yet, as bothersome and expensive as these problems are, they have not yet threatened national policy goals. However, if the networks themselves were seriously compromised, governments and the solvency of banks could be threatened.

Network operators' and users' claims that their security will soon be virtually impenetrable must be questioned. A technologically sophisticated Willy Sutton would view global financial networks with awe. After all, at least 80 percent of all money is now electronic, and those funds flow over global networks. If the funds could be hijacked or counterfeited as they were wired from account to account, the theft would beat bank robbery hands down.

The networks are protected by electronic safeguards, but if these defense systems can be beaten, huge sums of money could be siphoned off and laundered into hard currency. Indeed, in 1989 insiders were able to transfer $20 million from funds that did not exist at the Swiss Bank Corporation to the New York branch of the State Bank of New South Wales and then on to a desperate businessman in Australia who unwittingly handed over most of the money to the perpetrators. The thieves were caught and returned to Switzerland for prosecution, but only $8 million of the $20 million was recovered. This was a small case, but at least one study conducted by a former Federal Reserve official concluded that a large bank failure and default, perhaps instigated by fraud, could "generate a cascade of failures in other banks that were owed money by the defaulting institution" (cited in Peter Passell, Fast Money, *The New York Times Magazine*, 18 October 1992, 66, 75). National regulators working alone probably can do little to prevent such a debacle, but central bankers already are collaborating across national boundaries to avert such disasters.

Finally, governments also are wrestling with what they should do to safeguard individual privacy when everybody and everything is hooked together electronically. Some people like to be swamped with information, but many others despise being flooded with telephone solicitations, deluged by junk mail and catalogues, and overwhelmed by unwanted email messages. More troubling is the fact that everyone's personal life is becoming an open book. The electronic age is making individual spending patterns and preferences, credit records, house and mortgage records, bank accounts, and past brushes with the law available to any person, organization, or government bureaucrat who wants to check on them. It is legal in the United States to access public databases and to sell information collected about individuals. Blockbuster Video is said to have better records on many Mexican citizens than the Mexican government. The identification numbers of the AT&T Universal Cards allow data on spending records to be broken down further than ever before.[27] This information is valuable, and it is no accident that banks and credit card companies are the largest users of encryption technology to keep their databases secure.

In the early 1970s, some European countries and Canada passed laws to protect the privacy and records of citizens and firms. Privacy issues faded from view during the early 1980s but roared back on the political agenda a decade later. In the United States, the controversy over government plans to ensure that the FBI could tap phones and decipher certain coded messages with a court order brought personal privacy issues back to the forefront (James Fallows, Open Secrets, *The Atlantic Monthly*, June 1994, 46-50). To forestall harsh privacy protection laws, many companies do not sell or circulate personal data on their customers. Still, the mounting data banks on individuals, coupled with the ability of interested parties to tap that information, will likely make privacy protection a major issue. Some observers, such as Michael Nugent,

---

27. Or, a technologically sophisticated psychic could access extensive files available on line to demonstrate his psychic abilities to unsophisticated customers. Based on interviews by the author.

vice president and associate general counsel of Citicorp, contend that "the interplay among financial regulation, telecommunications regulations, and privacy regulation will determine the future of American banking overseas" (OTA 1992, 29).

Balancing Cooperative and Market Approaches

In complex environments with overlapping jurisdictions, policymakers and regulators are faced with difficult decisions about what they can do, what they ought to do, and how they ought to do it. The networked economy is global. Cyberspace is global. Governments will still intervene to protect their firms and telecommunications providers, but over time policies and regulations will increasingly be international.

One result of the rise of regional financial markets may be that smaller countries that are unable to insulate their markets may choose to link their fate formally to their neighbors'. The emergence of larger, more stable economic areas in developing and smaller industrial countries may help them attract external financing (Mytelka 1993). And, once begun, the logic of an effort such as monetary union in Europe impels participating countries to coordinate their regulatory and commercial policies at least in part to hold the project together. If Europe ends up with a common currency and some form of common central bank, then the United States, Japan, and Europe will also need to "agree on a general approach to exchange markets and to a domestic use of monetary and fiscal policies that can be interpreted by capital markets as stabilizing" (Axilrod 1990, 12).

New technologies also may promote multilateral policy coordination and regulatory cooperation. Technology innovation and the globalization of financial markets transformed the environment in which private banks and central banks operate, limiting the independence of domestic monetary and fiscal policies of even large countries. As capital markets become more integrated, the discipline of the market bears more heavily on wrong-headed government policies. These policies can be punished, as Mexico and, more recently, Southeast Asia discovered, by sharp depreciation of the exchange rate and drastic decline in share and property values (Quinn 1988, 35; OTA 1992, 35). Discipline of bad macro policy by market forces, abetted by the latest technology, can be applauded. But the same technology can create headaches when improperly used by those with larcenous intent.

To counterbalance this loss of control over private misbehavior, countries seek new policy and regulatory approaches. One possibility is that countries may adopt what French businessman Jean Saint-Geours called "co-regulation." Under co-regulation, authorities would cooperate to make their national regulations more transparent to each other, accept the right of other regulators to obtain information as requested, and in some instances adopt mutual recognition of each other's regulations (Saint-Geours 1991; 1992). Co-regulation could help harmonize international financial policies, but that is not enough.

As the integration of communications and financial technologies proceeds, supranational policymakers will need to determine the appropriate mix of co-regulation, cooperation, and harmonization. For example, they may work more closely together to craft macroeconomic and fiscal policies to safeguard against system failure. They may also act to minimize cross-border security risks involving payments and settlements. They will probably also try to settle problems of dual regulation in areas such as antitrust in global telecom networks.[28]

Ultimately, antitrust or competition policy could prove to be the critical regulatory tool in government arsenals. However, national antitrust policies still diverge dramatically, and significant harmonization will not proceed rapidly. Regulators also will need to manage jurisdictional overlaps involving financial and telecommunications issues.[29] At the same time, policymakers will need to recognize when to allow competition to substitute for policy intervention and regulation. When markets evolve toward a competitive equilibrium, even if there are significant winners and losers, the presumption should be against regulation.

At a minimum, regulators must prevent system failure. No one wants the system to collapse under the weight of massive financial flows. So, for the past two decades bank supervisors have worked to harmonize bank supervision and adopt stricter minimum standards. The Basel Committee on Banking Supervision recommended that both home and host countries give approval to banks that open offices in another country, that host country regulators routinely provide information to the home countries of banks operating within their borders, and that bank regulators share information among themselves (OTA 1992, 31). Each participating regulatory body agreed to monitor its financial institutions to ensure that they maintained adequate capital (Scott and Iwahara 1994; Kapstein 1994; Hale 1990). Governments forged these international agreements to ensure a global, level playing field, recognizing that the result

---

28. Indeed, antitrust policies are one way governments may respond to the rise of global networks. If governments develop general understandings about fair competition globally and begin to harmonize their individual policies, we may see the emergence of regional or global competition policy rules centered around telecommunications.

29. The proliferation of global networks and international strategic alliances force regulators to decide who has oversight powers. Problems of oversight once were left largely to the communications and finance ministries. Current developments raise questions about traditional regulatory practices, particularly about the division of authority between national regulators and international bodies such as the International Telecommunication Union (ITU), International Monetary Fund (IMF), and WTO. Under pressure from the WTO, which is trespassing into areas once reserved for the ITU, the ITU is slowly reforming itself. See Cowhey and Aronson (1991).

was a higher cost of capital for financial institutions. Most financial regulators now accept that problems cannot be solved by creating a new supranational regulatory body or by increasing the formal authority of the Basel Committee. The idea of a new international bureaucracy is neither efficient nor popular, and nobody wants to interfere directly with national sovereignty. Instead, regulators and bank supervisors need to co-regulate or harmonize their national supervisory standards to ensure that "(c)omplete and consistent information on national supervisory practices" is shared among countries thereby allowing national authorities to make better decisions on foreign bank operations within their jurisdiction (GAO 1994, 3).

Regulators also must contend with overlapping, contradictory, regulatory systems for finance and telecommunications. For instance, in the United States and Japan, banking and securities regulation are separated; elsewhere, universal banking is more common. With the merging of financial and telecommunications firms, countries are starting to harmonize their regulatory approaches internally and cross-nationally. Free market advocates favor liberalization and deregulation, warning that otherwise, as financial and telecommunications markets grow globally, business will migrate to less-regulated havens. At a minimum, regulatory simplification usually is better than increased regulatory oversight.[30]

Regulators may need to act when the system is out of kilter. But when confronted with the merger of communications and financial technologies, they should resist the urge to seek the order and predictability of the old regime. Regulators should not panic and overreach their capacity. Where competition exists and national and international antitrust policies are in place, regulators should keep their distance.

For instance, regulators need not increase their supervision of global card operations. To protect their customers, Visa and MasterCard already maintain extensive internal operating procedures with which their members must conform. Membership is extended only to qualified financial institutions under the laws of the home country. Most card associations are self-governing and self-policing in their efforts to minimize risk. Adding new global regulations, so long as self-policing is effective, would be redundant. Networking businesses that spring from the largely unregulated computer traditions also should not lightly be subjected to regulation. Just as the 1980s saw an explosion of competing firms, so the 1990s already have seen a proliferation of competing trading networks (Hoekman and Sauve 1991, 126). These developments will raise competitive opportunities but inevitably will be met with regulatory restraint. The

---

30. OTA (1990a, 71-72) argues that in the securities market, liquidity and country of origin considerations are more important than the level of regulation. Nor is it clear that investors prefer the least regulated markets.

pace of deregulation will ultimately depend on the success of governments in first promoting full and fair competition across national and international sectors and then stepping aside.

Walking the High Wire: Balancing Markets and Regulation

Technological change is transforming competition in communications and financial markets. The nonpolluting, job-creating industries that spring from these sectors offer great opportunities. As technology unfolds in the absence of global accords that guarantee open markets for services, regulators remain on call. Most regulators understand that to thrive the new industries will need room to grow. There are too many variables in play to allow government officials to fine tune their policies. They understand that economic micromanagement is inefficient. Often, market competition may prove a superior substitute for regulation, nationally and internationally.

But even as regulators proceed with caution, they still have a responsibility. Their challenge is to understand how and when, individually or in cooperation, they can promote competition and prevent collusion and monopoly. They need to decide when to allow markets to operate unencumbered, when to use what types of national regulation, and when to replace national regulation with international cooperation and supervision.

As the finance and information sectors grow more global and competitive, government regulatory approaches will need to evolve. Competitive markets, not regulators, ought for the most part to decide who should invest, how much, in which projects, and in which countries. In general, supply and demand, not governments, should dictate taste and content. If competitive global information and financial markets develop, greater economic prosperity and job creation are likely. Where oligopolists and protective national industrial policies triumph, the world will underperform its potential.

References

Aronson, Jonathan D. 1991. Communications, Diplomacy and International Relations. *Relazioni Internazionali* (March): 100-7.

Axilrod, Stephen. 1990. Interdependence of Capital Markets and Policy Implications. *Occasional Papers 32*. Washington: Group of Thirty.

Bank for International Settlements (BIS). 1989. *Payment Systems in Eleven Developed Countries*. Cited in OTA, US Banks and International Telecommunications. Chicago: Bank Administration Institute (May).

Beese, J. Carter. 1994. Bridges over Troubled Water: Patterns of Integration in and among International Financial Markets. *Project Promethee Perspectives* 22 (February): 4.

Capoglu, Gokhan. 1990. The Internationalization of Financial Markets and Competitiveness in the World Economy. *Journal of World Trade* 24, no. 2 (April): 114.

Cowhey, Peter, and Jonathan D. Aronson. 1989. Trade in Services and Changes in the World Telecommunications System. In *Changing Networks: Mexico's Telecommunications Options*, ed. by Peter F. Cowhey, Jonathan D. Aronson, and Gabriel Székely. La Jolla, CA: Center for US-Mexican Studies.

Cowhey, Peter, and Jonathan D. Aronson. 1991. The ITU in Transition. *Telecommunications Policy* 15 (4 August): 298-310.

Distler, Catherine. 1994. The Integration of Stock Exchanges in Europe. *Project Promethee Perspectives* 22 (February): 21-26.

General Accounting Office (GAO), US. 1994. International Banking: Strengthening the Framework for Supervising International Banks. *Report to Congressional Committees*. GAO/GGD-94-68. Washington: General Accounting Office (March).

Group of Thirty. 1989. Clearance and Settlement in the World's Securities Market. New York: Group of Thirty (March).

Hale, David D. 1990. Global Finance and the Retreat to Managed Trade. *Harvard Business Review* (January-February): 150-62.

Hinckley, Ronald H. 1989. Information Technology and Foreign Policy. *Institute for Information Studies 1989 Annual Report*. Queenstown, MD: The Aspen Institute.

Hoekman, Bernard M., and Pierre Sauve. 1991. Information Technology, Trade in Financial Services and Evolving Regulatory Priorities. In *Finance and the International Economy, The AMEX Bank Review Prize Essays, 5*, ed. by Richard O'Brien. London: Oxford University Press for the Amex Bank Review.

Kapstein, Ethan B. 1994. Governing Global Finance. *The Washington Quarterly* (Spring): 77-87.

Mytelka, Lynn Krieger. 1993. Rethinking Development: A Role for Innovation Networking in the "Other Two-thirds." *Futures*.

National Academy of Sciences. 1988. *Summary Report of a Workshop on The Revolution in Information and Communications Technology and the Conduct of U.S. Foreign Affairs*. Washington: National Academy Press.

Neal, Larry. 1990. *The Rise of Financial Capitalism: International Capital Markets in the Age of Reason*. New York: Cambridge University Press.

O'Brien, Richard. 1992. *Global Financial Integration: The End of Geography*. London: Pinter Publishers.

Office of Technology Assessment (OTA). 1990a. *Trading Around the Clock: Global Securities Markets and Information Technology-Background Paper*. OTA-BP-CIT-66. Washington: US Government Printing Office (July).

Office of Technology Assessment (OTA). 1990b. *Electronic Bulls & Bears: U.S. Securities Markets & Information Technology*, OTA-CIT-469. Washington: US Government Printing Office (September).

Office of Technology Assessment (OTA). 1992. U.S. Banks and International Telecommunications. Background Paper. OTA-BP-TCT-100. Washington: US Government Printing Office (September).

Quinn, James Brian. 1988. Technology in Services: Past Myths and Future Challenges. In *Technology in Services: Policies for Growth, Trade and Employment*, ed. by Bruce R. Guile and James Brian Quinn. Washington: National Academy of Engineering. Cited in OTA, US Banks and International Telecommunications.

Saint-Geours, Jean. 1991. Global Financial Networks and the Governance Imperative. *Project Promethee Perspectives* 17 (June): 3-6.

Saint-Geours, Jean. 1992. Networks and Markets: More Than a Marriage of Convenience. *Project Promethee Perspectives* 21 (December): 24-29.

Scott, Hal S., and Shinsaku Iwahara. 1994. In Search of a Level Playing Field: The Implementation of the Basle Capital Accord in Japan and the United States. *Occasional Paper 46*. Washington: Group of Thirty.

Spero, Joan. 1988-89. Guiding Global Finance. *Foreign Policy* 73 (Winter): 114-35.

Wriston, Walter. 1992. *The Twilight of Sovereignty: How the Information Revolution is Transforming*. New York: Scribners.

IV

CASE STUDY

8

Competition and Deregulation: An APEC Perspective

SHIN CHO AND MYEONGHO LEE

Introduction

Since the early 1980s, one country after another has dismantled the monopolistic provision of telecommunications. Industrialized countries took the first step, and many developing countries have now joined the liberalization parade. Member economies of the Asia Pacific Economic Cooperation (APEC) group generally support the telecom revolution, but they differ greatly in the scope of reforms. At one end of the scale, some APEC countries have introduced full-blown competition, privatization, and deregulation. At the opposite end, other APEC countries have allowed competition only in markets for customer premise equipment and enhanced services; they have not permitted private participation by domestic or foreign firms in the provision of basic network services. Moreover, plans for future telecommunications reform vary a great deal from country to country.

Telecom trade and investment initiatives have made headway in both the World Trade Organization (WTO) and APEC. Monopolized market structures, weak regulatory bodies, and restrictions on foreign ownership are the barriers now targeted in the steps toward full liberalization. Against this background, we examine the broad trends in telecommunications trade and reform in APEC and offer some policy recommendations.

Shin Cho and Myeongho Lee are senior research fellows at the Korea Information Society Development Institute, South Korea. The opinions expressed in this paper reflect the views of the authors and do not necessarily reflect those of the Korea Information Society Development Institute.

We begin by presenting an overview of the telecommunications sector in the APEC economies, describing the status of market reforms. Much greater detail, on a country-by-country basis, appears in appendix A. The next section of this chapter examines the liberalization process in greater detail, focusing on four issues: competition, privatization, restrictions on foreign ownership, and appropriate regulation. We conclude by discussing the possibilities of future cooperation among APEC member economies.

Overview of Telecommunications in APEC Economies

The Current State of Telecommunications in APEC

The Asian part of the Asia Pacific region has made remarkable economic progress in the last two decades, achieving growth rates considerably higher than other regions. Today, APEC economies as a whole contribute slightly more than half of world GDP. Trade volumes have steadily increased over the last decade, with intraregional trade accounting for two-thirds of the total. However, APEC consists of diverse member countries at vastly different stages of development, with dissimilar natural endowments, political histories, and economic characteristics (see table 8.1). Accommodating such differences, while achieving policy cooperation, represents a major challenge for APEC.

Economic growth fuels demand for telecommunications service from both residential and business users. But in order to sustain high overall growth rates, many APEC countries must significantly improve their telecommunications infrastructure. Within the APEC region, average "teledensity," the number of telephone lines per 100 inhabitants, rose from 10 in 1983 to 14 in 1993. Meanwhile, the quality of conventional services has been enhanced and new services have been introduced. Notably, mobile communications networks have expanded rapidly throughout the region. Countries with poor wire-line infrastructure have employed mobile communications as a substitute for a fixed-line telecommunications network, leapfrogging lengthy construction times and avoiding the investment required to build fixed networks. While these advances represent significant improvements, telecom service in many emerging Asian APEC countries remains far below the levels attained by industrialized APEC members (see table 8.2).

A recent statement published by the International Telecommunications Union (ITU) seemed to suggest that, with good intentions and sustained discipline, a country could choose between alternative organizational styles and still make major advances in its telecom sector:

> [R]emarkable network development can take place under different organizational circumstances and with countries at varying levels of social and

Table 8.1 Basic socioeconomic indicators for the APEC economies

	Population (millions) 1994	Population density (per square mile) 1994	GDP (billions of dollars) 1983	GDP (billions of dollars) 1993	GNP per capita (US dollars) 1983	GNP per capita (US dollars) 1993
Australia	17.8	2	153.4	280.9	9,966	15,905
Brunei	0.3	52	3.5	3.9	17,122	14,385
Canada	30.1	3	329.2	551.6	13,227	19,187
Chile	14.0	18	19.8	43.7	1,686	3,163
China	1,190.9	125	293.1	544.6	282	452
Hong Kong	5.8	5,461	28.4	108.5	5,300	18,521
Indonesia	189.9	99	85.4	138.5	546	730
Japan	124.8	330	1,186.3	4,215.5	9,943	33,764
South Korea	44.5	452	82.3	330.8	2,062	7,510
Malaysia	19.5	59	30.1	4.9	2,024	256
Mexico	92.9	47	148.9	490.1	1,994	5,373
New Zealand	3.5	13	20.6	42.3	6,450	12,224
Philippines	66.2	221	33.2	53.7	638	819
Singapore	2.8	4,545	17.4	55.1	7,213	19,194
Taiwan	21.3	592	52.4	223	2,818	10,630
Thailand	58.7	114	39.6	110	796	1,876
United States	259.8	28	3405	6,378	14,533	24,760
APEC	2,147.0	50	5,931	13,579	3,179	6,310
EU	370.0	111	2,328	8,694	6,459	23,539

economic development. Successful telecommunications development has taken place under government-run PTTs (Turkey), state-owned corporations with private sector management (Botswana), public corporations (South Korea), and private sector operators (Chile). Governments need to establish a clear policy about their role in this sector. The role boils down to essentially two choices: (1) an active and engaged operational role backed by financial support; or (2) a regulatory and policy-making role leaving operations to the private sector. Adopting and sticking to the choice—as shown by the high achievers—has resulted in sustained telecommunications development. (ITU 1994, 94)

While we accept the merit of allowing flexibility in the choice of development strategy, the lessons of the past two decades indicate that the market-oriented approach has been more rewarding for many APEC members.

Telecommunications Reform in APEC

Liberalization in the telecommunications sector typically includes four strands: competition, privatization, eased restrictions on foreign investors, and redesign of the regulatory structure. In addition to these, several features characterize telecom reform in APEC economies.

Table 8.2 Basic telecommunications in the APEC economies

	Main telephone lines (millions)		Teledensity (main lines per 100 persons)		Cellular phones (per 1000 persons)	
	1985	1995	1985	1995	1985	1995
Australia	6.2	9.2	39	51	0	128
Brunei	0.02	0.07	9	24	0	126
Canada	12.5	na	48	na	0	na
Chile	0.5	1.9	4	13	0	14
China	3.1	40.7	0	3	0	3
Hong Kong	1.8	3.3	32	53	1	123
Indonesia	0.6	3.3	0	2	0	1
Japan	45.3	61.0	38	49	1	81
South Korea	6.5	18.6	16	41	0	37
Malaysia	1.0	3.3	6	17	0	43
Mexico	3.7	8.8	5	10	0	7
New Zealand	1.3	1.7	40	48	0	108
Philippines	0.5	1.4	1	2	0	7
Singapore	0.8	1.4	31	47	0	98
Taiwan	4.2	9.2	22	43	0	36
Thailand	0.6	3.5	1	6	0	19
United States	118.3	165	49	63	1	128

na = not available.

First, telecommunications competition is no longer viewed as exceptional in Asia. Value-added services, such as electronic mail and voice mail, and customer premise equipment, such as private telephones and fax machines, are open to competition in virtually all APEC countries (see table 8.3). Competition has also spilled into wireless communications. Competitive long distance/international carrier operations have been licensed in about half of the APEC countries. Several countries allow competition in local services, but usually in a limited fashion.[1]

Second, we observe competition in markets for basic voice telephony regardless of the stage of telecom development. For many decades, it was assumed that competition was at odds with the goal of universal service. In fact, some monopoly telecoms have done a good job of providing universal service. For example, Taiwan has a well-developed

---

1. For example, the status of local services in the Philippines may be characterized as "partial competition," because local service providers in the Philippines are licensed to install networks mainly in regions that are not served by the incumbent carrier. Likewise, in China, while local exchange service is provided simultaneously by provincial governments and local governments, the system is not fully competitive because the purpose of simultaneous provision is not to stimulate competition between networks; rather, the government network is allowed to provide service only where the local government is unable to meet demand.

Table 8.3   Market structure of APEC telecommunications industry

	Local	Toll	International	Cellular	VAS[a]	CPE
Australia	C	C	C	C	C	C
Brunei	M	M	M	M	C	C
Canada	C[b]	C[b]	M	C	C	C
Chile	C	C	C	C	C	C
China	M	M	M	M	C	C
Hong Kong	C	C	M	C	C	C
Indonesia	M	M	C	C	C	C
Japan	C	C	C	C	C	C
South Korea	M	C	C	C	C	C
Malaysia	M	C	C	C	C	C
Mexico	M	M	M	C	C	C
New Zealand	C	C	C	C	C	C
Philippines	P	C	C	C	C	C
Singapore	M	M	M	C	C	C
Taiwan	M	M	M	M	C	C
Thailand	M	M	M	C	P	C
United States	C	C	C	C	C	C

C = competition   P = partial competition   M = monopoly

a. Value-added service (VAS) refers to telecommunications services with a computer-based information feature or performance that provides services such as electronic mail or voice mail.
b. Except in the Province of Saskatchewan where Sasktel, a provincial government corporation, provides these services.

telecommunications sector with about 40 main lines per 100 inhabitants, and this was built by a government monopoly with exclusive jurisdiction over wire-line voice telephony. More often, though, monopoly telecom has been in conflict with the goal of universal service. This is shown by the fact that emerging Asian countries generally have monopoly public telecom operators (PTOs) of long standing, and teledensity rates remain low—under 10 per 100 in most countries. Because of such statistics, the Philippines, a country with quite low teledensity, now has replaced its monopoly telecom system with one of the more liberal telecommunications sectors in Asia.

The third APEC characteristic is that the four strands of reform tend to be implemented together. Privatizing a state-run incumbent carrier is commonly accompanied by the introduction of competition, and often by allowing investment by foreign carriers as well. However, the specific elements and sequencing stages of reform differ from country to country.

Fourth, many countries have retained majority ownership of their PTOs. Among countries that have introduced competition in wire-line voice

Table 8.4   Status of privatization in APEC telecommunications industry

	Status of privatization
Australia	Telstra is government owned
	Sold AUSSAT, national satellite operator, to Optus in 1990
Brunei	Jabatan Telkom is government owned
Canada	Teleglobe privatized in 1987
	No government ownership in other carriers, except Sasktel
	Sasktel owned by the Province of Saskatchewan
Chile	CTC and ENTEL completely privatized in 1988
Hong Kong	No government ownership in Hong Kong Telecom
Indonesia	PT Telkom is government owned
	PT Indosat privatized in 1993, 35 percent sold
	PT Satelindo, 60 percent owned by private investors
Japan	NTT privatized since 1985, 65 percent owned by government
South Korea	Korea Telecom privatized since 1993, 20 percent sold
	Dacom completely privatized in 1993
	Korea Mobile Telecom completely privatized in 1994
Malaysia	Telekom Malaysia privatized in 1990, 25 percent sold
Mexico	Telmex completely privatized in 1994
New Zealand	TCNZ completely privatized in 1990
Philippines	Eight government operators serving rural areas (less than 10 percent of market)
Singapore	Singapore Telecom privatized in 1990, 10 percent sold
Taiwan	Part of ministry
Thailand	100 percent government-owned corporation
United States	100 percent private firms

telephony, for example, Australia and Indonesia still hold 100 percent ownership of the incumbent carriers. Japan, South Korea, Malaysia, and Singapore have sold 10 to 35 percent ownership of their PTOs, and thus still control their dominant incumbent carriers. Only three countries with public telecoms—Chile, Mexico, and New Zealand—have privatized a controlling percentage of shares. Of course, Canada and the United States have long had private monopoly telecoms, so the issue in these countries was competition, not privatization (see table 8.4).

Fifth, the mode of foreign participation often reflects the development stage of telecommunications (see table 8.5). In industrialized countries, foreign investors typically take over a privatized firm through a bidding process or through a public offering of stock. This is how British Telecom, for example, became a private company, and some emerging countries, such as Chile and Mexico, have followed the same route. Other emerging countries, however, have accepted foreign participation in the form of revenue sharing schemes. In these schemes, foreign carriers build networks, operate them jointly with the local public enterprise, share revenue over a specified period of time, and eventually transfer the facilities to the local partner. These build-operate-transfer (BOT) schemes

Table 8.5 Status of foreign ownership restriction in APEC

	Status of foreign ownership restriction
Australia	49 percent
Brunei	No ownership allowed
Canada	20 percent No voting shares allowed in Teleglobe
Chile	None
China	No ownership allowed
Hong Kong	None
Indonesia	No ownership allowed Foreign investors participate in joint operating scheme
Japan	33 percent 20 percent allowed in NTT and KDD
South Korea	33 percent No ownership allowed in telephone carriers (KT and Dacom)
Malaysia	25 percent
Mexico	49 percent
New Zealand	None A single foreigner cannot hold more than 49.9 percent in TCNZ
Philippines	40 percent
Singapore	49 percent 40 percent allowed in Singapore Telecom
Taiwan	No ownership allowed
Thailand	Foreign investors participate in BOT projects
Unites States	Case-by-case evaluation of telecom alliances with foreign firms 20 percent ownership for radio licenses

have been employed by several developing countries, including Indonesia and Thailand.

Another type of foreign participation in emerging countries is a joint venture or the provision of technical assistance. This type of foreign participation is especially common in areas that require sophisticated technology such as code division multiple access (CDMA) for mobile service; integrated service digital network (ISDN) for telephone, data, and video transmission; and electronic data interchange (EDI). Through joint venture and technical assistance, many emerging countries gain access to cutting-edge technology from industrialized countries.

Finally, government functions are often restructured along with privatization and competition. Historically, the government has had the dual functions of regulating and operating the telecommunications industry. However, APEC countries have increasingly vested policy and regulation functions in new pro-competitive bodies and have spun off service provision to independent entities. Some countries—Australia, Canada, Hong Kong, the Philippines, and the United States—have also separated the regulatory function from the policymaking function, and

Table 8.6 Restructuring of government functions in telecommunications

	Separation of operation/regulation	Separation of regulation/policy	Type of regulatory agency
Australia	yes	yes	semiautonomous
Brunei	no	no	ministry
Canada	yes	yes	semiautonomous
Chile	yes	no	ministry
China	no	no	ministry
Hong Kong	yes	yes	semiautonomous
Indonesia	yes	no	ministry
Japan	yes	no	ministry
South Korea	yes	no	ministry
Malaysia	yes	no	ministry
Mexico	yes	no	ministry
New Zealand	yes	no	ministry
Philippines	yes	yes	semiautonomous
Singapore	yes	no	ministry
Taiwan	no	no	ministry
Thailand	yes	no	ministry
United States	yes	yes	autonomous

semiautonomous or autonomous entities have been created to regulate the telecom industry (see table 8.6).

Liberalization Issues in Telecommunications in APEC

This section examines the four basic strands of telecommunications liberalization—competition, privatization, foreign investment, and appropriate regulation—as implemented in the APEC economies. Empirical testing gives some evidence of the effects of regulation and market structure.

Competition: Cause or Consequence?

As telecommunications reform has expanded to developing countries, the feasibility and desirability of competition in these countries have been debated. The conventional belief has been that competition is a consequence of development and that monopoly PTOs are necessary at the early stage of development. In particular, it was once widely believed that monopoly was required to facilitate cross-subsidization of subscribers who would otherwise remain out of the network. In this view, market mechanisms work efficiently only after the network has reached a substantial size and serves most of the potential subscribers.

In contrast, the new perspective challenges the concept that monopoly PTOs are best suited to expand telephone networks. Instead, competition is now seen as a promoter of teledensity (e.g., Harrington 1995). Advocates of the new paradigm ask a simple but powerful question: If the old paradigm was right, why do so many developing countries still have dismal teledensity rates, and why are they now voluntarily introducing competition?

Why Did Industrialized Countries Introduce Competition?

In recent years, technological advances have eroded the natural monopoly features of telecom networks. Digital technology has reduced the initial cost of investments required for new entrants to compete with incumbents. Alternative technologies, such as microwave and fiber optics, also make competition feasible at a smaller scale of investment. And, the economies of scale of the telecom industry disappear as the size of the telecommunications market grows: as the demand curve shifts out, it eventually intersects the average cost curve at the point where average cost increases, resulting in diseconomies of scale.[2]

The high cost of regulation is another important driver for the competitive model. Although regulation has demonstrated positive effects in preventing monopolistic pricing, numerous studies report that the regulation of monopoly carriers is typically accompanied by significant inefficiency.[3] Overall trends show that the symptom of market failure (natural monopoly) fades away, while that of government failure (cost of regulation) persists. In striking a balance between two failures, it might be quite natural to lean toward the market.

Effects of Competition in Industrialized Countries

Many studies have attempted to measure the effects of competition, especially in the United States, the United Kingdom, and Japan, where sufficient time has passed since competition was introduced. Most, if not all, studies offer support for the competitive mode. More specifically, competition is credited with bringing about the following effects:

■ Improvement in allocative efficiency, through rate rebalancing between local and long distance service;

■ Rapid reduction in overall rates, compared with monopolistic markets;

2. See Kiss and Lefebvre (1987), Roller (1990), and Shin and Ying (1992) for empirical examinations of economies of scale and natural monopoly features.

3. See Joskow and Rose (1989) for a comprehensive review of the effects of economic regulation.

- Improvement in performance, such as technical efficiency and productivity;

- Rapid introduction of new services, such as voice mail and internet;

- Increased demand for communications services owing to rate reductions and new and better services; and

- Higher investment levels.

The results of our empirical testing, as shown in table 8.7, illustrate that competition is strongly associated with advanced telecom infrastructure. In a monopoly situation, teledensity is reduced by 12 lines per 100 inhabitants. Further, competition seems to increase revenue per worker. Since telecom service is price elastic, lower rates actually increase revenue. These results indicate that breaking up a telecommunications monopoly improves both resource allocation and production efficiency.

As demonstrated by the example of rapidly growing internet users in the United States, lower connection charges create more internet demand. The initial high demand for internet connection motivated new providers to enter the server market, thereby creating a more competitive environment. As a result, connection charges dropped dramatically. This led to a new surge of new internet users. As in many other areas of telecom service, it is hard to determine whether high demand caused prices to drop or the drop in prices caused high demand. Whatever the causality, the introduction of competition seems to be a win-win game: consumers gain from lower prices and providers gain from higher revenue.

One of the harshest criticisms concerning the effects of US long distance competition comes from Huber, Kellog, and Thorne (1993). They argue that competition in the US long distance market was "propped up by regulation, and operating under an AT&T supplied canopy of umbrella pricing" (Huber, Kellog, and Thorne 1993, 352). In other words, this was "hothouse competition," not a Darwinian struggle for survival. The authors acknowledge, however, that going back to monopoly in the United States would involve costly regulation, which would be even worse. As they point out, "the one thing worse than the illusion of efficient competition may be the illusion of efficient regulation" (Huber, Kellog, and Thorne 1993, 352).

Local Service Competition in Developing Countries

The most contentious dimension of competition is the local telephone market, which was the background for both landmark legislation in the United States (the US Telecommunications Act of 1996) and the WTO Agreement on Basic Telecommunications, concluded in February 1997. In countries with teledensity rates over 30, allowing a second or third

Table 8.7 Effects of regulation on telecommunications infrastructure, 1990-95 (dependent variable: log [revenue per main line[a]])

Independent variables	Telecom-developing country group (teledensity[b]< 30)		Telecom-developed country group (teledensity > 30)	
	Coefficient	T-value	Coefficient	T-value
Monopoly[c]	-5,667	2.8	-24,743	1.8
Separation[d]	-19,396	15.3	22,805	3.3
Privatization[e]	218	4.9	150	1.2
Foreign ownership[f]	-60	6.2	-497	2.9
Log (number of mobile phone subscribers[g])	2,147	7.9	-1,602	1.8
Log (per capita GDP[h] in US$)	1,757	1.4	3,014	1.3

Note: The data encompasses 12 countries: Australia, Canada, China, Hong Kong, Indonesia, Japan, South Korea, Mexico, New Zealand, the Philippines, Singapore, and the United States.

a. Revenue per main line is an indirect measure of the capital efficiency of the telecom industry. It is derived by dividing total telecom revenue by the number of main lines (in thousands).

b. Teledensity refers to the number of main lines connecting subscribers to switchboards and is calculated per 100 inhabitants. The sample is divided into two groups: telecom-developed countries, which have teledensity rates of 30 or higher, and telecom-developing countries, which have teledensity rates of 30 or lower. This separation enables a differentiated analysis of the effects of regulation, which may vary depending on the level of development.

c. Monopoly indicates the degree of competitiveness of the telecom industry. To measure the degree of competitiveness, the Herfindahl index is used. The Herfindahl index is a market concentration ratio, which is calculated as

$$H = \sum_{i=1}^{n} \alpha_i^2,$$

where n = number of telecom firms, and α_i denotes firm i's market share. If one monopoly firm dominates the telecom industry, the Herfindahl index is one. The Herfindahl index asymptotically decreases to zero as the market becomes more competitive.

d. Separation is a dummy variable. If the regulatory body is separated from telecom operations, separation has a value of one. Otherwise, it has a value of zero.

e. Privatization refers to the percentage of the dominant telecom firm's shares held by private investors. For example, if the dominant telecom firm is fully privatized, privatization is 100. If the government holds 100 percent of the dominant telecom firm's stake, privatization is zero.

f. Foreign ownership refers to the percentage of the dominant telecom firm's shares held by foreigners. For example, if the dominant telecom firm is established jointly, and the government holds 50 percent while foreign firms hold 50 percent, foreign ownership is 50.

g. Number of mobile phone subscribers is the total of analog and digital mobile phone subscribers expressed in millions.

h. Per capita GDP represents real GDP expressed in US dollars.

local carrier implies either that additional local loops will be installed—whether wireless or wire—or that the incumbent carrier will be required to offer new entrants reasonable access on its network. In either case, competition means offering existing subscribers alternative carriers.

In the case of developing countries with low teledensity, competition usually means building new infrastructure to offer service to households not previously reached by the incumbent carrier and facilitating interconnection to the incumbent network. The threat of entry by new competitors thus spurs existing carriers to expand their networks as fast as possible to secure lucrative regions. They have less incentive to enter regions where another carrier has already established a network. As a result, local competition in developing countries generally leads to rapid expansion of the telephone network rather than the massive duplication of facilities. To be sure, as in industrialized countries, competition may also bring new telephone lines to metropolitan areas where the demand for services is high. This will not contribute to the expansion of network, but it may very well reduce rates and enhance services in the metropolitan areas.

An important limitation to the competitive model comes from the goal of ensuring universal service. Traditionally, government PTOs have used long distance revenues to cross-subsidize local service in an effort to promote universal service. Competition in long distance markets reduces rates in that sector, making it difficult to continue cross-subsidies and, in the view of some PTOs, threatening the goal of universal service. However, an ITU report (1994) found that almost 50 million people in the world are on waiting lists for telephone service, with an average waiting time of more than 18 months. This huge waiting list indicates the ineffectiveness of cross-subsidy policy for reaching the goal of universal service. Many households are willing to pay the competitive price—they need a telephone line—not a subsidy. Indeed, recent experience in China and the Philippines indicates that competition is more successful than cross-subsidies in expanding the local network and stimulating the growth of telephone service.[4]

Some governments have erected variants of the wide-open competitive model: granting exclusive rights to different carriers to install local networks in each region. This is supposed to ensure service even in less profitable regions; it is also meant to provide "yardsticks" for comparing different carriers. Other countries have tried build-operate-transfer (BOT) plans, with foreign carriers entering long-term partnerships with the incumbent. Neither of these approaches provides meaningful competition in the provision of local services, though both may represent an improvement over a nationwide PTO monopoly.

---

4. See Bornholz and Evans (1983), Smith and Staple (1994), and Ure (1995).

Competition in Other Telecommunications Markets

In telecommunications markets for other kinds of service—for example, long distance, mobile, and value-added service—the arguments against competition never had much persuasive force. Most APEC countries have already introduced competition in mobile and value-added services; the restrictions on competition that remain in long distance and international services are essentially justified as a funding source for cross-subsidizing local service. If and when local service markets are resolved in favor of the competitive model and do not require subsidies, the argument against competition in other markets will fall away.

Privatization: Improving PTO Performance

Economists generally believe that a public enterprise is inherently less efficient than a private firm, and that the ownership structure of a public enterprise is mainly to blame. By definition, in a public enterprise there is no monitoring of management by private shareholders, and the government "owners" may have little incentive to ensure efficiency or even expand the network (De Alessi 1974).

Given these characteristics of public ownership, privatization is widely suggested as a natural solution (Vickers and Yarrow 1988). In cases where public ownership is justified by market failure, the common view among economists holds that the public enterprise should be privatized as market failure diminishes and as competition is introduced. In sum, privatization of incumbent carriers is considered part of the process of introducing competition.

Directions for Privatization

Since liberalization of the telecommunications sector usually requires several policy measures, many countries follow a staged approach and implement the changes sequentially over an extended period of time. Setting the priority among policy measures then becomes an issue. The sequencing order between competition and privatization is especially controversial.

Studies indicate that privatization of the incumbent without competition from new carriers may improve production efficiency, but lower costs may not be passed on to telecom users in the form of lower prices. Instead, the fruits of greater efficiency are claimed by the privatized incumbent as increased profits. Only competition from new entrants reliably results in both productivity gains and allocative efficiency through lower prices. Table 8.8 illustrates this point more clearly. Privatization of PTOs in low-teledensity countries increases revenue per worker. Our findings indicate that if 10 percent more shares of PTOs are held by

Table 8.8 Effects of regulation on efficiency of telecom firms, 1990-95
(Dependent variable: revenue growth per worker[a])

Independent variables	General		Telecom-developed country group (teledensity[b] > 30)		Telecom-developing country group (teledensity < 30)	
	Coefficients	T-value	Coefficients	T-value	Coefficients	T-value
Monopoly[c]	-153	2.5	-106	2.8	-91	4.2
Separation[d]	-6	0.2	50	2.7	143	10.3
Privatization[e]	-1	2.0	2	0.4	1	2.5
Foreign ownership[f]	1	1.9	0	1.6	2	2.2
Log (number of mobile phone subscribers[g])	-4	1.1	-17	7.1	16	5.3
Log (per capita GDP[h] in US$)	17	1.6	40	5.8	33	2.4

Note: The data encompasses 12 countries: Australia, Canada, China, Hong Kong, Indonesia, Japan, South Korea, Mexico, New Zealand, the Philippines, Singapore, and the United States.

a. Revenue growth per worker is derived by dividing total telecom revenue expressed in US dollars by the number of full-time workers in the telecom industry. Revenue per worker indirectly measures the labor efficiency of the telecom industry.

b. Teledensity refers to the number of main lines connecting subscribers to switchboards and is calculated per 100 inhabitants. The sample is divided into two groups: telecom-developed countries, which have teledensity rates of 30 or higher, and telecom-developing countries, which have teledensity rates of 30 or lower. This separation enables a differentiated analysis of the effects of regulation, which may vary depending on the level of development.

c. Monopoly indicates the degree of competitiveness of the telecom industry. To measure the degree of competitiveness, the Herfindahl index is used. The Herfindahl index is a market concentration ratio, which is calculated as

$$H = \sum_{i=1}^{n} \alpha_i^2,$$

where n = number of telecom firms, and α_i denotes firm i's market share. If one monopoly firm dominates the telecom industry, the Herfindahl index is one. The Herfindahl index asymptotically decreases to zero as the market becomes more competitive.

d. Separation is a dummy variable. If the regulatory body is separated from telecom operations, separation has a value of one. Otherwise, it has a value of zero.

e. Privatization refers to the percentage of the dominant telecom firm's shares held by private investors. For example, if the dominant telecom firm is fully privatized, privatization is 100. If the government holds 100 percent of the dominant telecom firm's stake, privatization is zero.

f. Foreign ownership refers to the percentage of the dominant telecom firm's shares held by foreigners. For example, if the dominant telecom firm is established jointly, and the government holds 50 percent while foreign firms hold 50 percent, foreign ownership is 50.

g. Number of mobile phone subscribers is the total of analog and digital mobile phone subscribers expressed in millions.

h. Per capita GDP represents real GDP expressed in US dollars.

private investors, revenue per worker increases by $35,500 per year, holding other variables constant. However, in high-teledensity countries, privatization decreases revenue per worker although the effect is not statistically significant. These results may indicate that, in low-teledensity countries, privatized but noncompeting PTOs substantially reduce the size of the workforce and may charge higher service rates. Both changes increase revenue per worker. In high-teledensity countries where competition is usually more intense, additional privatization may have less impact on workforce efficiency and service rates.

The main practical reason why privatization has sometimes taken priority over competition was to "package" the PTO so that it could be sold for more money as a dominant or monopoly carrier. For this reason, both Mexico and Singapore gave privatization priority over other forms of liberalization. The incumbents were granted legal monopolies until 1996 and 2007, respectively. By contrast, other APEC countries that introduced privatization and competition at approximately the same time were not so concerned about maximizing the proceeds from selling their PTOs. With or without competition, privatization is one of the key drivers to improving telecom infrastructure. The statistical analysis suggests that low-teledensity countries can increase their teledensity rate by 30 lines per 100 inhabitants if the dominant telecom operator is fully privatized (see table 8.7).

A related policy issue is whether to privatize completely or partially. Many APEC countries sold less than half the shares in the PTO and retained control of management. Chile, Mexico, and New Zealand are the only countries that simply handed control to the private sector. The main reason other countries have retained control is to guard against abrupt changes in employment practices (e.g., laying off thousands of workers) or in rate structures (e.g., sharply raising local service rates). Whatever the rationale, the full benefits of privatization can be expected only after private investors control the majority of shares.

Market Access of Foreign Carriers: Efficiency or Sovereignty?

It was not until the early 1990s that allowing market access to foreign telecom investors became an important issue. Foreign ownership reflects two kinds of incentives: the incentive of carriers in industrialized countries to enlarge their markets and provide global end-to-end service, and the incentive of developing countries to finance the expansion of their networks. However, some developing countries have tried to limit the role of foreign investment to a lender status, through BOT or technical assistance schemes, rather than an owner status, with full control.

The statistical analysis in table 8.7 suggests that foreign ownership is associated with lower teledensity. However, this negative effect may not reflect the effects of foreign ownership so much as the preexisting condition of a country's telecom development. A poor developing country, which has a relatively weak telecom infrastructure, is more likely to permit foreign ownership in its PTO because the country needs both capital and technology. On the other hand, a relatively rich developing country is more likely to limit foreign participation to technological assistance.

The governments and carriers in the newly industrialized economies (NIEs) (notably, South Korea, Taiwan, Hong Kong, and Singapore) have different incentives than the developing countries. They have good infrastructure and they are in the process of introducing competition. The NIE carriers themselves want to reach outside their geographic boundaries to tap new business opportunities. Unlike carriers in developing countries, they do not need foreign investment to finance network expansion. At the same time, the NIEs fear that investment access means that their domestic markets will be dominated by foreign carriers, and that the NIE carriers will lose the chance to stay competitive both at home and in international markets. The situation among the NIEs varies somewhat according to their size. The small countries, Hong Kong and Singapore, have a relatively larger stake outside their boundaries and therefore are more willing to open their domestic markets to foreign carriers in an effort to become international telecom hubs. The larger NIEs, South Korea and Taiwan, are more conservative about opening their domestic markets.

Foreign Ownership and Sovereignty

In summary, reservations remain over foreign participation in the telecommunications sector, although most observers agree that it is inevitable. The WTO Agreement on Basic Telecommunications, concluded in February 1997, represents a major step toward unlimited foreign ownership, but much remains to be implemented and many countries are doubtful. Are their reservations based on ungrounded fears, or is there really something "bad" about foreign ownership?

From the standpoint of managerial efficiency, best-practice technology, and rapid network expansion, it is hard to challenge unfettered foreign ownership. Moreover, foreign ownership plays a key role by providing financial resources, without which it might be virtually impossible to expand the network.[5] But not all policymakers are principally

---

5. One of the common characteristics of "high achievers" in telecommunications development is that they reinvested a large portion (typically more than 50 percent) of their earnings for the expansion of the network, whereas in some developing countries investment is below 25 percent of earnings. One reason that telecommunications revenues are not reinvested is that governments use these resources for other activities (ITU 1994). In

concerned about efficiency and forefront technology. Instead, they invoke national sovereignty arguments against market opening.[6] The telecommunications industry is considered a strategic sector in economic and social development (akin to steel a generation ago). This provides the rationale for the government to ensure that key decisions are made at home, by either top public officials or local companies. These decisions may concern cross-subsidy policy, the regional priority of network expansion, universal service goals, or the public purchase of equipment from indigenous manufacturers. Concerns about national control may make little sense to economists; however, such concerns decide the outcome between permitting or preventing foreign ownership in many developing countries.

China, for example, welcomed foreign participation in building urban networks, especially in mobile and other value-added services. Foreign participants brought more sophisticated technologies, which cut the cost of building the network. However, for national security reasons, China's regulatory body, the Ministry of Posts and Telecommunications (MPT), assigned the foreign firms operation in particular regions—Siemens in Beijing, NEC in Tianjin, and Alcatel in Shanghai. In this way, the MPT prevented individual foreign firms from acquiring excessive power in China.

Future Directions for Market Access

For most industrialized countries, investment access is a good thing, because domestic carriers are competitive and will expand on a global basis. Foreign ownership within the industrialized countries typically occurs as a consequence of strategic alliances, as in the cases of BT-MCI (which led to a takeover) and DBP Telekom-France Telecom-Sprint.

For the NIEs, allowing foreign ownership could provide an opportunity to promote efficiency and competitiveness. Permitting one or more foreign carriers in the market intensifies competition; to survive the market test, domestic incumbents must streamline. Foreign ownership can also facilitate partnerships with the major international carriers. Therefore, lifting most restrictions on foreign ownership seems desirable for the NIEs.

For the developing countries, however, sovereignty issues are sometimes decisive. In such cases, BOT schemes at the local level may be the

---

this case, they will run short of capital for the expansion of the telecommunications sector and will have to rely on outside resources unless the governments readjust their priorities in investment.

6. For this discussion about sovereignty, we relied mainly on Globerman (1995), Joseph (1995), and Preston (1995).

most practical tool for expanding the telephone network. Limited share-holding by foreign carriers in local service networks may also be politically acceptable. In other sectors, especially mobile and value-added services, outright foreign ownership may be less sensitive. In some large countries, like China and Indonesia, full-scale competition and foreign ownership may be allowed in all sectors.

Regulation to Ensure Market Performance

Countries that introduced competition in the early 1980s discovered that *allowing* competition is only a prelude to *achieving* competition. Even after the entry of new carriers, the existence of a dominant carrier might stifle competition. Without active regulatory oversight, the market might regress to de facto monopoly or cartel.

Regulatory initiatives in APEC countries have generally been limited to removing entry barriers. The regulatory initiatives needed to ensure competition are slow to be put in place. The WTO Agreement on Basic Telecommunications could spur pro-competitive regulation. This was one of its major objectives and is reflected in the umbrella principles.

Competition Safeguards

Anticompetitive practices typically stem from the fact that the incumbent carrier controls a vertically integrated network and monopolizes the local loop. With this power, incumbents have been known to use revenue from sheltered markets for predatory pricing; to create technical barriers to interconnection; to discriminate in the terms and conditions of access to the local loop among new entrants; to conceal their own network and subscriber information base; and to abuse the information obtained on subscribers to other carriers.[7]

To curtail these potential anticompetitive practices, various competition safeguards are required. Compelling separate accounts between different service elements; enforcing equal access for all new entrants; requiring transparent access charge plans; and ensuring open network architecture (ONA) are all typical components of competition safeguards. Nonetheless, many countries have yet to adopt these safeguards. And even if they are adopted, there will be many problems in implementation and dispute settlement. In other words, the achievement of competition remains at some distance even after the decision has been made to allow competition.

---

7. For a comprehensive discussion of potential anticompetitive behavior by vertically integrated carriers, see Brennan (1995) and Noll (1995).

Price Deregulation

Telecom price regulation is still widely practiced among APEC countries. Many countries seem unaware that price deregulation should follow entry deregulation. The data compiled by APEC (1993; 1994) show that all the member countries, except New Zealand, are involved in setting telecom tariffs in one way or another.[8] Only Australia and the United States introduced substantial price deregulation by setting price ceilings for the dominant carriers and no tariff regulations on the new carriers. Reflecting the WTO Agreement on Basic Telecommunications, many other countries are replacing detailed tariffs with price ceilings. One of the highlights of the WTO agreement is the commitment by 65 countries to ensure pro-competitive regulatory principles. These countries will find it easier to accomplish this commitment if they deregulate prices.

Restructuring Government Functions

Restructured government functions are typically included in a telecom reform package. At the very least, it would look awkward if the government ministry continued to run the PTO while regulating private carriers at the same time; conflicts of interest would surely arise. As a first step, therefore, operating functions are usually separated from the ministry.

In most APEC countries, the ministry still remains in charge of telecom policy and regulatory functions. Ministry-style regulation may work effectively to control a monopolistic carrier; however, with the introduction of competition the focus of regulation drastically changes from setting detailed rates to implementing competitive safeguards and settling disputes among service providers. To handle such responsibilities, the regulatory agency should have regulatory expertise, a transparent process, and first-stage judicial power. Ministry-style regulations do not seem a good fit for these requirements.

Cooperation in APEC

The discussion so far sets out our position that liberalization in telecommunications will contribute to the growth and efficiency not only of the industry but also of the whole economy. As liberalization proceeds, international cooperation will become a more serious issue. Differences in market structures, regulatory environments, and restrictions on foreign

---

8. Even the New Zealand government sets limits on price increases; i.e., the standard residential rate for a phone line may not rise faster than the consumer price index.

ownership can become sources of conflict among APEC members. In addition, as international telecommunications traffic increases by leaps and bounds, a number of policy issues will arise. Current debates on the reform of accounting rates and on the regulation of international callback services are prominent examples.

The role of APEC in the telecommunications sector could become more important as the level of telecom interdependence increases. Given the great diversity among APEC members, nobody believes that APEC can become an integrated entity like the European Union. However, at their 1993 meeting in Seattle, APEC leaders embraced the vision of an "APEC Economic Community." And at their 1994 meeting in Bogor, they adopted the goal of liberalizing regional trade and investment by the year 2010 for developed members and by 2020 for developing members. The Bogor Declaration provided momentum for APEC to discuss substantive issues in telecommunications, such as improving market access.

In line with the Bogor spirit, participants in the APEC Ministerial Meeting on Telecommunications and Information Industry, held in Seoul in 1995, agreed to "promote telecommunications cooperation, trade and investment liberalization and facilitation in the region, and enhance the information infrastructure in the Asia Pacific region through an Asia Pacific Information Infrastructure (APII) initiative" (APEC Ministerial Meeting 1995a). The ministers adopted 10 principles for the APII:

- Encouraging member economies in the construction of domestic telecommunications and information infrastructure based on their own reality;

- Promoting a competition-driven environment;

- Encouraging business/private-sector investment and participation;

- Creating a flexible policy and regulatory framework;

- Intensifying cooperation among member economies;

- Narrowing the infrastructure gap between the advanced and developing economies;

- Ensuring open and nondiscriminatory access to public telecommunications networks for all information providers and users in accordance with domestic laws and regulations;

- Ensuring universal provision of and access to public telecommunications services;

- Promoting diversity of content, including cultural and linguistic diversity; and

- Ensuring the protection of intellectual property rights, privacy, and data security.

To promote the APII principles, the APEC Telecommunications Working Group is supposed to

■ Define the commitments necessary to achieve free and open trade and investment by the 2010/2020 Bogor time frame;

■ Coordinate steps and milestones with target dates defined by each member; and

■ Identify areas where collective commitments can be made by consensus (APEC Ministerial Meeting 1995b).

This is a tall order, and it would be reasonable to say that APEC has not yet gone beyond preliminary discussions of policy and regulatory issues in the telecom sector. So far, the APEC Telecommunications Working Group has developed two sets of guidelines to improve access to telecom equipment and services: (1) Guidelines for Regional Harmonization of Equipment Certification, and (2) Guidelines for Trade in International Value-Added Network Services. Not until the 13th Working Group Meeting, held in March 1996, did the Group establish a Liberalization Steering Group to discuss trade and investment liberalization issues.

As an interim agenda for the APII initiative, our recommendation is that the APEC Working Group develop guidelines on specific issues such as

■ International simple resale and international callback service;

■ Accounting rate reform;

■ Spectrum assignment and licensing for Low Earth Orbit Satellite Systems;

■ Standardization in EDI, ISDN, Cellular/PCS, and other services and equipment;[9] and

■ Enhanced market access to countries that are not yet participants in the WTO (notably, China, Taiwan, and Vietnam).

Setting guidelines on these issues would contribute to the Bogor goal of free regional trade and investment. Moreover, agreement among APEC

---

9. Electronic data interchange (EDI) is the electronic exchange of trading documents that precludes the need for traditional paper documentation. Integrated service digital network (ISDN) is a network in which the same switches establish a simultaneous interface for data, video, telex, and facsimile. Personal communications service (PCS) is a type of communication in which a user carries a small communications unit that can be reached regardless of location. The system has advanced digital features that allow a high degree of customization.

members on guidelines for some of these specific issues could pave the way for global understandings under WTO and ITU auspices. Given the great diversity of APEC member economies, a certain degree of flexibility with regard to policy options should be honored in the process toward these objectives.

References

Asia Pacific Economic Cooperation (APEC). 1993. *The State of Telecommunications Infrastructure and Regulatory Environment of APEC Economies*, vol. 1.

Asia Pacific Economic Cooperation (APEC). 1994. *The State of Telecommunications Infrastructure and Regulatory Environment of APEC Economies*, vol. 2.

Asia Pacific Economic Cooperation (APEC) Ministerial Meeting. 1995a. Seoul Declaration for the Asia Pacific Information Infrastructure. APEC Ministerial Meeting on Telecommunications and Information Structure, Seoul (29-30 May).

Asia Pacific Economic Cooperation (APEC) Ministerial Meeting. 1995b. Action Plan to Promote the APII Arising Out of APEC Ministerial Meeting on Telecommunications and Information Industry. APEC Ministerial Meeting on Telecommunications and Information Structure, Seoul (29-30 May).

Bornholz, R., and D. Evans. 1983. The Early History of Competition in the Telephone Industry. In *Breaking Up Bell: Essays on Industrial Organization and Regulation*, ed. by D. Evans. New York: North-Holland.

Brennan, T. 1995. Is the Theory behind US v. AT&T Applicable Today? *Antitrust Bulletin* 40: 455-82.

De Alessi, L. 1974. An Economic Analysis of Government Ownership and Regulation. *Public Choice* 14: 1-42.

Globerman, S. 1995. Foreign Ownership in Telecommunications: A Policy Perspective. *Telecommunications Policy* 19: 21-28.

Harrington, A. 1995. Companies and Capital in Asia-Pacific Telecommunications. In *Telecommunications in Asia: Policy, Planning and Development*, ed. by J. Ure. Hong Kong: Hong Kong University Press.

Huber, P., M. Kellog, and J. Thorne. 1993. *The Geodesic Network II: 1993 Report on Competition in the Telephone Industry*.

International Telecommunications Union (ITU). 1994. *World Telecommunications Development Report: World Telecommunications Indicators*. Geneva: ITU.

Joseph, R. 1995. Direct Foreign Investment in Telecommunications: A Review of Attitudes in Australia, New Zealand, France, Germany and the UK. *Telecommunications Policy* 19: 413-26.

Joskow, P., and N. Rose. 1989. The Effects of Economic Regulation. In *Handbook of Industrial Organization*, ed. by R. Schmalensee and R. Willig. New York: North-Holland.

Kiss, F., and J. Lefebvre. 1987. Econometric Models of Telecommunications Firms: A Survey. *Revue Economique, Special Issue of Telecommunications Economics*: 307-74.

Noll, R. 1995. The Role of Antitrust in Telecommunications. *Antitrust Bulletin* 40: 501-28.

Preston, P. 1995. Competition in the Telecommunications Infrastructure: Implications for the Peripheral Regions and Small Countries of Europe. *Telecommunications Policy* 19: 253-71.

Roller, L. 1990. Modeling Cost Structure: The Bell System Revisited. *Applied Economics* 22: 1661-74.

Shin, R., and J. Ying. 1992. Unnatural Monopolies in Local Telephone. *Rand Journal of Economics* 23: 171-83.

Smith, P., and G. Staple. 1994. *Telecommunications Sector Reform in Asia: Toward a New Pragmatism.* World Bank Discussion Paper No. 232. Washington: World Bank.

Spulber, D. 1995. Deregulating Telecommunications. *Yale Journal on Regulation* 12: 25-68.

Ure, J. ed. 1995. *Telecommunications in Asia: Policy, Planning and Development.* Hong Kong: Hong Kong University Press.

Vickers, J., and G. Yarrow. 1988. *Privatization: An Economic Analysis.* Cambridge, MA: MIT Press.

APPENDIX

Telecommunications Services in the Asia Pacific Countries

ERIKA WADA AND TOMOHIKO ASANO

Overview

The telecommunications industries in the Asia Pacific Economic Cooperation (APEC) countries have grown rapidly. In the last decade, the number of main telephone lines in APEC economies grew by an average of 224 percent; international telephone traffic grew by 820 percent; and productivity (measured by revenue per worker) grew by 150 percent. Telecommunications expansion was greatest in China, where main line growth was 1,205 percent. Figures A1.1 through A1.5 show further measures of telecommunications expansion in APEC.

This remarkable growth was noted at the May 1995 APEC ministerial meeting held in Seoul, South Korea. The meeting focused on the further development of Asian Pacific telecommunications infrastructure, which many leaders believe is crucial to achieving the goals of the Bogor Declaration (free trade by 2020 for developing countries and 2010 for developed countries). Furthermore, each APEC member country realizes the value of APEC-wide telecommunications for its own economic growth. These factors have been a major force for improving the infrastructure of APEC telecommunications.

As infrastructure has improved, the price of telecommunications service has plummeted. For example, prices of basic telecommunications

Erika Wada is and Tomohiko Asano was a research assistant at the Institute for International Economics.

Figure A1.1 Teledensity, 1995 (per 100 people)

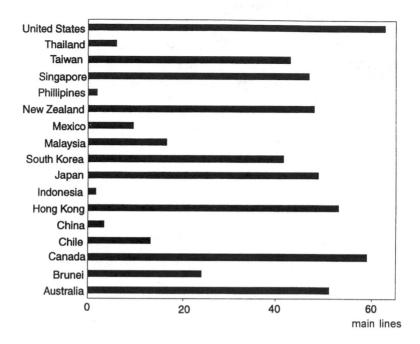

main lines

Figure A1.2 Main telephone line growth, 1985-95

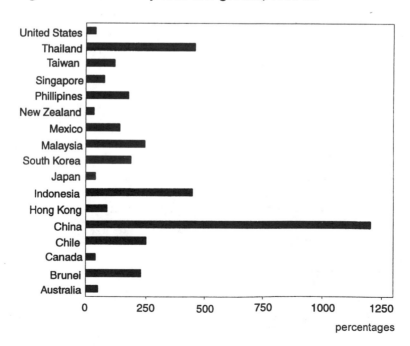

percentages

Figure A1.3 Growth of international traffic, 1985-95

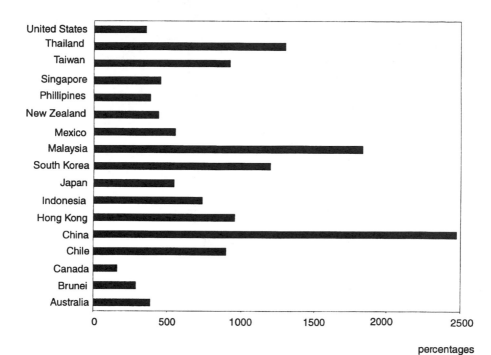

percentages

Figure A1.4 Growth of total telecom revenue, 1985-95

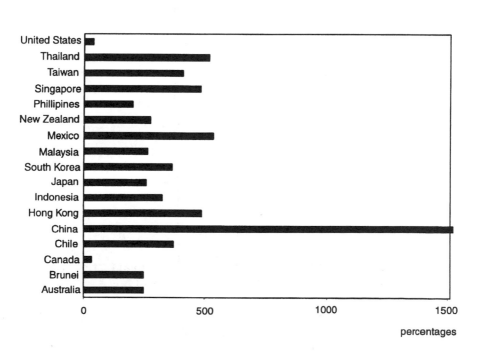

percentages

Figure A1.5 Growth of telecommunications revenue per worker 1985-95

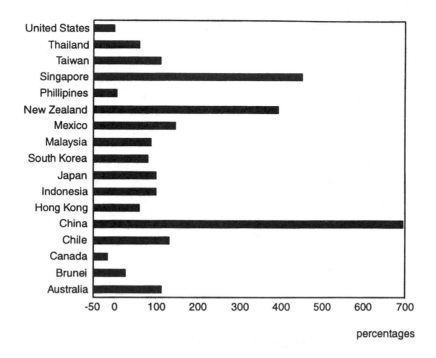

percentages

services dropped by about 14 percent from 1996 to 1997 (see table A1.1). The growing popularity of cellular phones have reinforced a declining cost structure for handsets and subscription charges. Furthermore, the rapid growth of cellular has resulted in a supply shortage, and foreign cellular operators have been invited by governments to build high capacity network systems in many metropolitan areas. Meanwhile, internet users are multiplying daily, circulating information on everything from fashion trends to investment options.

The country reports document globalization trends in APEC. They also illustrate the steps taken in various APEC member countries to bring competition to an industry that was once dominated by monopoly firms.

Australia

Highlights from offer at World Trade Organization (WTO) Negotiations on Basic Telecommunications (WTO 1997):

■ Offers unrestricted competition in virtually all basic telecommunications services as of July 1997.

Table A1.1 Cost comparison of telecom services[a]

Country	Business[a] Price 1997 (US$)	Rate of change 1996-1997 (percentages)	Home[b] Price 1997 (US$)	Rate of change 1996-1997 (percentages)
Australia	4,234	-11	2,333	-20
China	3,790	-23	2,616	-23
Hong Kong	3,393	-9	1,992	-19
India	6,356	29	4,327	2
Indonesia	4,728	-6	2,549	-22
Japan	6,587	-19	3,736	-11
Malaysia	4,060	-5	2,347	-13
New Zealand	7,004	4	3,684	0
Philippines	4,000	5	2,313	14
Singapore	2,894	-31	1,720	-39
South Korea	3,538	-26	2,281	-21
Taiwan	4,040	-23	2,045	-27
Thailand	4,172	-7	2,692	-10
United Kingdom	5,045	-37	2,485	na
United States[b]	4,994	-25	2,065	na
Vietnam	4,960	-21	2,702	-25
Simple average	4612	-13	2618	-15

na = not available.

a. The business basket includes:
Fixed-line service: connection and rental of single fixed-wire business line, including any other one-time charges; 2,000 minutes a month of local calls and one 10-minute IDD call every two months to each of the other 15 countries.
Cellular-service: purchase of Ericsson 388 hand sets or equivalent, connection and subscription for single GSM or equivalent; 700 minutes a month of local calls, including free minutes.

b. The home basket includes:
Fixed-line service: connection and rental of single fixed-wire residential line; 1,000 minutes a month of local calls and one 10-minute IDD call every three months to each of the other 15 countries.
Cellular-service: purchase of Ericsson 388 hand sets or equivalent, connection and subscription for single GSM or equivalent; 200 minutes a month of local calls, including free minutes.

Source: Asian Wall Street Journal, 10 June 1997.

- Commits to existing free markets for voice telephone and many other basic services on a resale basis.

- Offers to end limits on the number of satellite service providers (currently set at two) and on primary suppliers of public mobile cellular telephony and facilities-based carriers (both currently set at three) as of July 1997.

- Offers no limits on foreign equity for new carriers.

- Permits foreign equity up to 11.7 percent of the government-controlled carrier, Telstra, and requires majority Australian ownership of the mobile carrier, Vodafone.

- Removes foreign equity limitation on Optus.

- Commits to the Reference Paper on regulatory principles.

Infrastructure and Policy

Basic telecommunications infrastructure is well developed in Australia; as of 1995, 9.2 million main telephone lines, or more than 50 per 100 people, were installed, and 95 percent of households were covered. This telecommunications infrastructure was supported by a large, regulated telecommunications industry. For the past decade, reforming the telecommunications industry to allow open competition has been at the center of the national political debate (along with reforms in other key regulated service sectors such as finance and civil aviation).

Before the telecommunications deregulation movement was launched in the late 1980s, three publicly owned carriers dominated the entire Australian telecommunications market. Australian Telecommunications Commission (Telecom), a licensed domestic communications monopoly, accounted for about 85 percent of the industry's total operating revenue in 1990. Overseas Telecommunications Commission (OTC) offered international service and accounted for 14 percent of the total industry revenue in 1990. Because OTC was not permitted to run its own domestic network, it had to interconnect its international facilities with Telecom's domestic network. Aussat, which began the national satellite operation in the late 1970s, entered commercial service in 1985. Originally, Aussat was allowed to use its satellite facilities only to provide broadcast services—it was not allowed to provide basic telephone services until 1992. By law, each of these carriers complemented rather than competed with the others.

Telecommunications reform gradually took shape in the late 1980s. During the third term of then-Prime Minister Robert J. L. Hawke's Labor government (1987-90), an economic recession prompted the government to think about revitalizing the nation's international competitiveness of national industries by making strategic reforms in key industrial sectors. Inevitably, privatizing major government-owned companies in important service sectors—notably Telecom, Qantas (aviation), and Commonwealth Bank (finance)—became a main topic of debate.

The Telecommunications Act of 1989 established the Australian Telecommunications Authority (Austel) as a comprehensive and independent regulatory authority. The policymakers did not address the monopoly issue. Instead, they gave Austel responsibility for monitoring the monopoly of telecommunications providers to protect consumers without bringing competition into the market. Austel introduced a price-cap system[1] that led to a decline in service rates.

The next reform step was taken in 1990, when the government announced its plan to introduce full competition by July 1997 and to permit oligopolistic competition until that date. Under the Telecommunications Act of 1991, Telecom and OTC were merged to create a full-range service provider, Australian and Overseas Telecommunications (later Telstra). To create a telecommunications duopoly, Optus Communications was licensed as a second full-service carrier in November 1991. Optus purchased Aussat, which had been unprofitable for years, from the government for A$800 million and, in 1992, began providing mobile phone and long distance services. In June 1996, Optus launched local phone service through its joint-venture subsidiary, Optus Vision, thus becoming a full-service competitor to Telstra. In 1993, Vodafone became the third entrant in the market, providing mobile communications.

Today, Telstra remains owned by the federal government and continues to dominate local, long distance, and international telephone service in Australia. The country's powerful trade unions and the Australian Labor Party (ALP) caucus have opposed privatization. In the run-up to the 1993 general election, Prime Minister Paul Keating pledged to retain Telstra as a publicly owned enterprise. But private-sector involvement was seen as almost inevitable if Telstra was to remain at the forefront of advanced communications technology. Consequently, there was talk of floating off parts of Telstra's business or bringing in joint-venture partners on a limited basis.

In 1996, John Howard's Liberal-National coalition gained control of the government, making privatization of Telstra conceivable. The coalition plans to sell one-third of Telstra's shares, estimated to be worth A$8 billion. The coalition contemplates limiting foreign ownership of this one-third to 35 percent. The plan was approved by Parliament in December 1996, and the share offer—the largest in Australia's history—was to be floated in the second half of 1997. Meanwhile, in accordance with the earlier schedule, a fully competitive market was to be introduced in 1997.

Local and Long Distance Services

While the Australian telecommunications market has come a long way toward becoming fully competitive. In 1992, the long distance market was the first to be opened to competition when the two new entrants, Optus

---

1. Under the system, rate changes were restricted below an annual CPI rate (Australian CPI was 5.5 percent in 1994) minus a pre-determined margin, which was 1.8 percent. This yields the price cap of 3.7 percent.

Table A2.1 Revenue and share in revenue of general service providers, Australia, 1991-96

Provider	1991-92	1992-93	1993-94	1994-95	1995-96
Revenue (millions of A$)					
Telstra	12,229	12,656	13,636	14,081	14,800[a]
Optus	55	300	841	1,435	1,994
Vodafone	na	na	9	33	135
Share in revenue (percentages)					
Telstra	99.6	97.7	94.1	90.6	87.4[a]
Optus	0.4	2.3	5.8	9.2	11.8
Vodafone	na	na	0.1	0.2	0.8

na = not available.

Note: Australia's population in 1995 was 18.1 million.

a. Estimated.

Sources: Telstra annual report 1995; Optus 1996; Vodafone: <http://www.vodaphone.co.uk> (21 November 1997).

and Vodafone, challenged the dominant player, Telstra, in several markets. A 51-percent share of Optus is owned by domestic investors, including Mayne Nickless (25 percent), AMP Society (10.3 percent), and AIDC Fund (9.6 percent). Foreign carriers hold the remaining 49 percent: Cable & Wireless of the United Kingdom holds 24.5 percent and BellSouth of the United States, 24.5 percent. Vodafone is a part of Vodafone Group of the United Kingdom (which owns 95 percent of its shares).

In 1992, when Optus entered, Telstra held 99.6 percent of the total telecommunications revenue (see table A2.1). By 1995, Optus and Vodafone claimed 11.8 percent and 0.8 percent, respectively, of the total telecommunications revenue. Optus has already gained 27.8 percent of the market share in long distance service. Optus also has a significant share of the international market; in 1994, it transmitted 21.2 percent of outgoing international call minutes, while Telstra transmitted 71.6 percent. Optus's success may have be attributable in part to two procompetition policies Austel had in place during the transition period. First, Austel imposed a low interconnection rate on Telstra that will be applied as long as Telstra maintains its dominant position. Second, it conducted a series of ballots letting customers choose their primary long distance company.

Since its start in 1993, Vodafone has been providing fully digitized mobile phone service based on Global System for Mobile Telecommunications (GSM). Its network covered 80 percent of Australia's population in 1996. In 1996, when the local market was opened, Optus Vision, a cable TV operator jointly owned by Optus and Continental Cablevision of the United States, rolled out local call and data transmission services through

Table A2.2   Telecommunications price records, Australia, 1990-95

Service	1990-91	1991-92	1992-93	1993-94	1994-95
Average standard price (A$)					
Local	0.21	0.22	0.25	0.25	0.25
Long distance[a]	1.16	1.16	1.10	1.08	1.08
New connection fee	240.00	240.00	210.00	210.00	210.00
Average standard price index (1990 = 100)					
Local	100	105	119	119	119
Long distance[a]	100	100	95	93	93
New connection fee	100	100	88	88	88
CPI	100	103	104	106	108

a. Cost of a call from Melbourne-Sydney for 3 minutes during the day.

Source: <http://www.Austel.gov.au> (1996).

Cablevision's cable network. Optus Vision immediately offered a competitive local call rate of 20 cents, some 20 percent cheaper than Telstra's rate.

Rate changes during the transition period were not drastic, at least in terms of official standard rates (table A2.2). However, because of the price cap system and the budding forces of competition, real rates, after taking discount programs into account, have been declining. According to recent statistics (table A2.3), in 1994, Telstra's overall service rate declined by 5.6 percent. In terms of service categories, long distance rates fell 9 percent, international service rates declined 8.7 percent, and local service rates dropped 2 percent. The features of the price cap system were largely reflected differences between categories.

Mobile Communications Services

The mobile communications market has grown steadily since its inception in 1987. By 1994, the mobile penetration rate in Australia reached 12.8 percent, one of the highest in the world. Then, in 1995, the number of subscribers increased by another 70 percent.

Competition intensified in 1993, when Optus began the resale of Telstra's analog circuits. Meanwhile, Optus expanded its own digital network, and Vodafone started operations using a fully digitized GSM network. As of 1995, Optus and Vodafone had market shares of 31.5 percent and 5.3 percent, respectively. Facing competition with the new entrants, Telstra cut its tariff for mobile communications services by 7.1 percent in 1994 and 9.7 percent in 1995. Rates for mobile service declined more than long distance or international rates.

"Digitization" remains an issue in Australia. In a plan designed to improve the quality of service and foster advanced telecommunications

Table A2.3   Telstra services rates data, Australia, 1993-95 (percentages)

Service	Revenue-weighted change in rates, 1993-95			Price cap[a] 1994-95
	Standard	Discount	Total	
Local	-0.49	-3.36	-3.85	1.82
Long distance	-0.43	-14.02	-14.45	-3.68
International	-3.38	-8.65	-12.03	-3.68
Mobile	-2.58	-14.23	-16.81	none
Total	-0.52	-8.79	-9.31	-4.54

a. Price cap only applies to 1994-95.

Source: Competitive Safeguards & Carrier Performance 1995, Austel (1996).

technology, the federal government and the operators have agreed to phase out analog services and replace them with digital services by 2000. In spite of the phase-out plan, however, the number of analog service subscribers continues to rise, because the service rates are more attractive (see table A2.4).

Other Services

Value-added services were opened to full competition in 1989.[2]

1997 Environment

Table A2.5 summarizes Australia's environment for telecommunications services in 1997.

Canada

Highlights from offer at WTO Negotiations on Basic Telecommunications (WTO 1997):

■ Certain routing restrictions and foreign equity limits on many services are scheduled to be phased out by the year 2000,

---

2. The basic sources of information on Australia are Australian Government Industry Commission, *Annual Report 1990-91*, Australian Government Publishing Service, and *Financial Times*, 25 February 1994.

Table A2.4 Mobile subscribers, Australia, 1993-96 (thousands)

Type	1993-94	1994-95	1995-96
Analog	1,185	1,996	2,675
Telstra	786	1,387	1,880
Optus	399	609	795
Digital	35	309	1,250
Telstra	na	114	600
Optus	na	114	440
Vodafone	na	81	210
Total	1,220	2,305	3,925
(percent change)	60.5	88.9	70.3

Share comparison (percentage)

	1995-96
By type	
Analog	68.2
Digital	31.8
By carrier	
Telstra	63.2
Optus	31.5
Vodafone	5.4

na = not available.

Source: <http://www.austel.gov.au> (1997).

Table A2.5 Telecommunications services environment, Australia, 1997

Type of service	Price (A$)
Monthly charge, residential fixed line	11.65
Installation fee, residential fixed line	173.00
Local call per 3 minutes	0.25
IDD[a] call to New York per minute	1.40
Cellular monthly charge, standard plan	35.00
Installation wait for a residential line	7 days

a. IDD (International Direct Dial) is the cost of a phone call to New York using an incumbent carrier, Telstra.

Source: *Asian Wall Street Journal*, 10 June 1997.

and Telesat's exclusive rights on satellite facilities and earth stations serving the US/Canada market are to be eliminated in March 2000.

- Removes a requirement that Canadian equity holdings in mobile satellite systems must equal Canadian usage levels.

- Agrees to remove, as of October 1998, restrictions on obtaining licenses to land submarine cables.

- Offers a liberalized regime for resale-based competition in local telephone services and in most other basic telecommunications services.

- Limits foreign equity in all facilities-based suppliers to 20 percent direct and 46.7 percent combined direct and indirect foreign ownership.

- Teleglobe's monopoly on overseas (non-US) facilities-based service will be eliminated, and its foreign equity limits will be raised to 46.7 percent in October 1998.

- Maintains a few limitations on market access for telephone service in certain cities or provinces.

- Commits to the Reference Paper on regulatory principles.

Infrastructure and Policy

Telecommunications infrastructure in Canada is well developed; in 1994, the country had 17.3 million installed main telephone lines, of which 92 percent were digitized. At 59 main lines per 100 inhabitants, Canada has the second-highest teledensity among APEC members, just below the 60 per 100 in the United States. The share of telecommunications in national investment has generally been above 3 percent (see table A3.1).

Many areas of telecommunications service within Canada have been liberalized in terms of regulatory barriers. Important exceptions are satellites and international calls. However, regulatory liberalization does not necessarily mean that a competitive market environment has been established. This is partly because the Canadian telecommunications industry is structured to remain practically intact.

For most of this century, telecommunications service in each province was provided by a single private or government operator that connected its network to a long distance carrier shared with other local operators. In 1931, local operators formed an alliance, Trans-Canada Telephone System, later renamed Stentor Canadian Network Management (Stentor). The Stentor group now consists of 11 provincial local operators: BC TEL, Bell

Table A3.1 Telecommunications investment, Canada

	1985	1990	1991	1992	1993	1994	1995	1996
Investment								
All industries (C$)	na	na	128,010	122,188	121,253	130,131	127,956	126,009
Telecoms services (C$)	na	na	4,972	5,328	4,686	4,334	4,075	4,275
Share in total (percentages)	na	na	3.88	4.36	3.86	3.33	3.18	3.39
Employment								
All industries (thousands)	9,651	11,146	10,549	10,246	10,271	10,447	10,673	na
Telecoms services (thousands)	109	124	122	119	106	109	114	na
Share in total (percentages)	1.13	1.12	1.16	1.16	1.03	1.04	1.07	na

na = not available.

Source: <http://www.statcan.ca/english/pgdb/economy.htm> (1997).

Canada, Island Tel, Manitoba Telephone System (MTS), Maritime Tel & Tel (MT&T), NewTel Communications, NBTel, NorthwesTel, Québec-Téléphone, Sask-Tel, and TELUS (table A3.2 provides information about each of these operators). In 1992, the group formed Stentor Resource Center as a policy advisory organization and Stentor Telecom Policy as an engineering and marketing organization.

In response to rapid technological advance and intense international competition, the Canadian government has adopted promarket policies. It took its first major step toward telecommunications liberalization in 1992, when the Canadian Radio-Television and Telecommunications Commission (CRTC), the national regulatory authority, issued Decision 92-12, which opened up the long distance market to new entrants outside the Stentor alliance. Following this reform, Unitel Communications launched its long distance service. In 1994, CRTC issued Decision 94-19 to introduce competition in the local service market.

Even after access barriers were eliminated, CRTC exercised its regulatory power in two ways: (1) a rate-of-return system under which it permits the operators to charge tariffs yielding no more than a "reasonable" return, and (2) price controls on local telephone service. Before 1992, local service was financially supported by cross-subsidies from long distance operations; however, the cross-subsidies have not been viable since the long distance call markets were liberalized.

In Decision 94-19, CRTC announced that by 1998 it would introduce a price cap system to replace the rate-of-return system. During the transitional period, from 1994 to 1997, CRTC is easing price regulations on local service. To do this, CRTC has divided telecommunications service activities into two categories: (1) utility, which consists of local and interconnection access services, and (2) competitive, which includes everything else. After January 1995, CRTC applied the rate-of-return system

Table A3.2 Stentor companies and main statistics, Canada, 1985 and 1995

	Bell Canada	BC TEL	Island Tel	MTS	MT&T	NBTel	NewTel	Northwes-Tel	Québec-Téléphone	SaskTel	TELUS	Telesat	Total
1985													
Total revenues (mil C$)	5,5	1,211	33	343	303	244	136	na	177	424	1,054	101	9,591
Share (percentage)	58.0	12.6	0.3	3.6	3.2	2.6	1.4	na	1.8	4.4	11.0	1.1	100
Total route length (km)	62,446	11,186	328	9,249	3,605	2,968	5,336	na	4,188	17,069	9,333	na	125,708
Fiber optics (km)	2,318	170	na	58	89	na	21	na	375	3,455	na	na	6,486
Microwave (km)	60,128	11,016	328	9,191	3,516	2,968	5,315	na	3,813	13,614	9,333	na	119,222
Lines (thousands)	7,092	1,525	53	517	385	304	148	na	221	509	915	na	11,671
Share (percentage)	60.8	13.1	0.5	4.4	3.3	2.6	1.3	na	1.9	4.4	7.8	na	100
1995													
Total revenues (mil C$)	8,183	2,391	62	540	463	372	262	115	266	539	1,211	242	14,646
Share(percentage)	55.9	16.3	0.4	3.7	3.2	2.5	1.8	0.8	1.8	3.7	8.3	1.7	100
Total route length (km)	57,186	26,132	588	16,623	4,743	4,266	3,314	8,415	7,165	12,941	41,691	na	183,064
Fiber optics (km)	32,171	14,000	588	10,700	4,743	3,515	3,314	na	3,783	12,941	8,072	na	93,827
Microwave (km)	25,015	12,132	na	5,923	na	751	na	8,415	3,382		33,619	na	89,237
Lines (thousands)	10,001	2,335	84	864	552	527	279	73	281	595	1,277	na	16,870
Share (percentage)	59.3	13.8	0.5	5.1	3.3	3.1	1.7	0.4	1.7	3.5	7.6	na	100

na = not available.

Source: <http://www.stentor.ca/bodyl.cfm?page_id=stentor_alliance.html> (21 November 1997).

to utility activities only. Also, to reduce losses from local services, CRTC allowed local operators to raise the monthly local service charge by C$2 in each year from 1996 to 1998.

Throughout the history of Canadian telecommunications, private companies have operated alongside government companies. Some government operators have been privatized in the last decade. In 1987, Teleglobe Canada, the international communications service monopoly, was sold to Teleglobe (formerly Memotec Data). At the same time, Teleglobe was given a license as the exclusive provider of international telecommunications services for the next five years. This approval was renewed in 1992 for another five years. BCE, the holding company that owns Bell Canada, also owns some Teleglobe shares.

Another telecommunications monopoly that was privatized was Telesat Canada, the exclusive provider of satellite communications services, which had been jointly owned by the federal government and provincial carriers, including some Stentor members. In 1992, the government sold its share to Alouette Telecommunications, 90 percent of which is owned by Stentor members and 10 percent by Spar Aerospace.

Through 1997, Canada limits foreign ownership to 25 percent in facilities-based telecommunications common carriers and 33 percent in telecommunications holding companies (notably BCE).

Local and Long Distance Services

Reflecting the policy shift toward liberalization, competition has been introduced into the market. As long ago as 1983, Unitel filed an application to offer long distance services. Not until June 1992 did the government finally decide to open the long distance market to competition and allow Unitel to offer long distance switched service. Following deregulation in 1993, Sprint Canada, wholly owned by Call-Net Enterprises, also began offering facilities-based long distance services.

Both new entrants in the long distance market have strategic alliances with US operators. Sprint of the United States owns 25 percent of the equity stake in Call-Net Enterprises, the parent firm of Sprint Canada. AT&T previously owned 22.5 percent of Unitel's shares, while Canada's largest cable television (CATV, community antenna television) company, Rogers Communications, and a Canadian railway company, Canadian Pacific Limited (CP), held 48 percent and 29.5 percent, respectively. After Unitel experienced financial difficulty in 1994, AT&T, the Bank of Nova Scotia, the Toronto-Dominion Bank, and the Royal Bank of Canada agreed to take control from Rogers and CP. Unitel plans to change its name to AT&T Canada Long Distance Services.

In addition to Sprint Canada and AT&T Canada, two other companies —a British Columbia operator, Westel Telecommunications, and a former regional reseller, fONOROLA—are operating facilities-based long distance communications services.

Restrictions on the resale of long distance services have been eased. Resellers are permitted to use wide-area telephone service (WATS) to provide competitive long distance service, without having to build a network. A small but growing portion of the market for public and private long distance service is held by resellers.

In the local telephone service market, competition is still limited. The Stentor members basically have local monopolies. Major CATV companies have offered private local access services, but they have less than 1 percent of the market. Once the current debate on convergence between the TV and telecom industries is settled, the telephone companies and broadcasting companies are expected to enter each other's markets. This development should stimulate competition in the local call market.

Mobile Communications Services

In 1994, Canada had 2.56 million mobile subscribers, a penetration rate of 8.7 percent. One consortium of regional providers and one national operator are competing in the mobile communications market. Mobility Canada is a consortium of regional cellular service operators, each of which is licensed to manage mobile communications in its service area (Mobility Canada's structure is similar to Stentor's; in fact, most members of Mobility Canada are Stentor member companies). Mobility Canada holds more than 80 percent of market share in revenue terms and has about 70 percent of total subscribers. The other competitor in this market, Rogers Cantel Mobile Communications, is a national operator providing services in all provinces.

As a part of its Information Highway plan, the Canadian government plans to introduce a new digitized system, Personal Communication Services (PCS), using the 2 Ghz range (existing cellular systems operate below the 1 Ghz range). In December 1995, licenses to operate nationwide service were awarded to Clearnet PCS and Microcell Network. Licenses for more limited service areas were awarded to Roger Cantel Mobile and Mobility Personacom Canada.

China

Infrastructure and Policy

During the 1980s, the telecommunications industry became one of the target industries for China's construction push. As a result, between

1981 and 1995 the number of main lines installed was expanded fivefold to 40.7 billion, or 3 lines per 100 inhabitants. Telecommunications infrastructure development has begun in urban areas; in 1994, Guangdong province had 16 telephones per 100 people, far exceeding the national average. Uneven development is not, however, a main concern in China.

By the year 2000, China plans to increase the number of telephone lines to 120 million, serving 30 percent of urban and 5 percent of rural dwellers, at a cost of about US$50 billion.[3] The additional network capacity would be equivalent to three times the existing network system in the United Kingdom.

As part of this ambitious plan, in March 1997, the Ministry of Posts and Telecommunications (MPT), the central regulatory body, announced a US$720 million increase in its investment in the provinces of Shandong and Sichuan. In Shandong, one million new telephone users will be connected to the system; and in Sichuan, 3,500 kilometers of optical cables will be laid, increasing the switching capacity of one million lines. The national network is also expanding: in northwest China, 2,000 km of optical-fiber line, connecting Beijing, Hohhot, Yinchuan, and Lanzhou, began operating in 1997.

The MPT regulates and develops the Chinese telecommunications industry. There are some 30 regional Post and Telecommunications Administrations (P&Ts) under the MPT umbrella. Since the late 1970s, the P&Ts' accounting systems have been independent of the MPT (Tan 1994). This accounting separation allows the P&Ts to use part of their revenues for investment and to seek financial support from local governments and other local interests.

Since the 1980s, China's local exchange capacity has been constructed simultaneously by regional P&Ts and township and village enterprises (TVEs). P&Ts are entitled to install facilities throughout a province. However, in smaller communities, where waiting lists are long and service is limited, TVEs are permitted to install their own local exchange facilities and interconnect them on standard terms with the public switched telephone network. Currently, over 200 TVEs and more than 2,000 mining, oil-field, and industrial areas without P&T service have developed their own networks. In some provinces, TVEs now account for 30 to 40 percent of the total number of exchange lines. This system has indirectly undermined the MPT monopoly.

An increase in the number of communications satellites and an extension of earth station networks to all provincial capitals are key priorities

---

3. The five-year plan announced in May 1994 calls for 16 fiber-optic trunk cables crossing the country, connecting a network of mobile phone systems (*The Financial Times*, 17 May 1994).

for the government. By 1995, China had seven communications satellites, including two Chinese-made systems.

Foreign firms are strictly prohibited from operating telecommunications networks. The MPT says that restrictions on foreign involvement will be gradually removed from value-added services: e-mail, voice mail, data interchange, videotext, and online database systems. However, the MPT still restricts foreign ownership of local, regional, and national networks. No plans have been set to relax these restrictions.

Despite the MPT's efforts, foreign participation in China's telecommunications industry seems likely. Foreign participation first began in the telecommunications equipment market. Alcatel of France started its production of central switching equipment in the early 1980s and now has about 40 percent of the market.[4] In 1979, AT&T was invited to build a joint-venture factory in China to produce switches. This would have been the first such plant; however, US export controls prevented AT&T from meeting China's request to provide the requisite level of technology (Warwick 1994). In the mid-1980s, the United States relaxed its controls on exports to China, and AT&T established a fiber-optic joint venture in 1985. In 1997, AT&T signed a memorandum of understanding that contemplated an electronic data interchange network (EDIN) in Guangdong. Other major foreign participants—Ericsson of Sweden, NEC of Japan, Siemens of Germany, and Motorola of the United States—have focused on value-added and mobile services.

Local and Long Distance Services

Until recently, the MPT provided almost all basic telephone services. The only organizations permitted to run private networks were other government ministries. The Ministries of Aerospace, Transport (roads, rivers, and canals), Energy Resources (oil fields, mines, and electricity grid), and Railways, as well as the People's Liberation Army, have been authorized to own private networks (both fixed-wire and radio-based facilities) since 1976. By 1993, those alternative networks accounted for about 40 percent of China's total capacity.

Some of these ministries moved toward the creation of a second nationwide telecommunications network, called China United Telecommunications (Unicom). In July 1994, Unicom was licensed to offer full telecommunications service in competition with the MPT. Simultaneously, the MPT's operating and regulatory divisions were separated to make it

---

4. Among public switching networks in China, Sino-foreign joint ventures account for more than 40 percent, while foreign companies supply about 50 percent. Purely domestic companies scrabble for about 7 percent of the market (*The Journal of Commerce*, 3 January 1995).

Table A4.1 Numbers of phone subscriptions, China, 1991-97

Year	Fixed line (millions)	Cellular (thousands)
1991	6.9	18.3
1992	8.5	47.5
1993	11.7	176.9
1994	17.0	638.0
1995	27.0	1,568.0
1996	40.7	3,629.0
1997	na	6,000.0

na = not available.

Sources: International Telecommunication Union, 1995 World Telecommunication Development Report; Asian Wall Street Journal, 10 June 1997.

easier for the new operators to establish themselves (*The Financial Times*, 21 July 1994).

Three ministries—Electronic Industries, Electric Power, and Railways—became major players in the telecommunications industry, because they possess extensive networks that enable them to offer service under the new Unicom network. The China International Trust & Investment Corporation (CITIC) and the People's Liberation Army are also trying to enter China's lucrative telecommunications market. Although foreign equity participation is not permitted, companies like Nynex and GTE of the United States have engaged in advisory roles.

Meanwhile, officials at the local level appear ready to welcome foreign participation. In late 1993, the mayor of Shanghai told a group of western telecommunications operators that he would like to arrange a pilot project in which foreign firms (perhaps working together in a consortium) would operate a new network in Shanghai. Other local officials have also called for a relaxation of MPT regulations.

Mobile Communications Services

Because the connection charge for a new telephone line is high in China, and the waiting list is long, buying a mobile phone is the cheapest and quickest way to get telephone service. Wireless technology can provide service at about half the cost of conventional networks. The MPT estimated that by the end of 1994 there would be two million cellular telephone users in 400 cities (concentrated, however, in places like booming Guangdong province) and 15 million pager users. Indeed, China is the third largest mobile market, after the United States and Japan (see table A4.1).

In late October 1993, the government announced that it would strengthen its control over telecommunications. The main target was ancillary services such as paging, mobile phones, satellite systems, and electronic data interchange networks. According to Chinese officials, these services are "chaotic," implying that MPT's monopoly in these service areas is eroding. In Beijing, for example, some 70 private paging firms are operating. The government therefore announced that all cordless telephones, pagers, and similar electronic devices must be registered, all ventures must seek new government approval, and limits on foreign ownership would remain in effect in the telecommunications industry. The official rationale was national security, not economics. In fact, the government was concerned about losing its grip over information flows within China and between China and the rest of the world. The announcement surprised many foreign investors.

Despite this tighter government control over mobile phone service, the movement toward competition and foreign participation continued. For example, JiTong Communications, a domestic firm established in 1993 and controlled by the Ministry of Electronics Industries, has obtained approval to establish pager, cordless telephone, and data communications services. Champion Technology of Hong Kong holds a 40-percent stake in a network in Sichuan province. In October 1994, Hong Kong Telecom, majority owned by Cable & Wireless of the United Kingdom, began to cooperate with Beijing's regional telecommunications administration to upgrade Beijing's cellular network. This deal heightened interest among other western telecommunications operators, particularly US companies.

In February 1997, the Directorate-General of Post (DGP) signed a distribution agreement with Ericsson under which Ericsson mobile phones will be distributed in post offices throughout China. Indeed, China (including Hong Kong) has become Ericsson's second most important market after the United States. Motorola has contracted with Beijing Telecom to provide base stations for its GSM, a digital network. Motorola's network system capacity will increase the number of subscribers to 500,000 from the current 250,000 and will cover 230 cities using MPT's mobile phone system. In May 1997, Siemens of Germany also started a GSM network operation covering most parts of the province of Qinghai.

In urban areas such as Beijing, mobile phones are a necessity, and existing networks are unable to meet the high demand. In April 1997, the MPT announced that it would encourage foreign participation to facilitate the transfer of advanced technologies, especially the code division multiple access (CDMA) technology. Lucent Technology of the United States first launched CDMA networks in China in the spring of 1997. Several trial CDMA networks are now being built in Guangzhou, Beijing, and Shanghai.

Table A4.2 Telecommunications services environment, China, 1997

Type of service	Price (US$)
Monthly charge, residential fixed line	2.89
Installation fee, residential fixed line	421.95
Local calls per 3 minutes	0.01
IDD call to New York per minute[a]	2.22
Cellular monthly charge, standard plan	6.03
Installation wait for residential line	60 days

a. IDD (International Direct Dial) is the cost of a phone call to New York, using the MPT.

Source: Asian Wall Street Journal, 10 June 1997.

Other Services

The Chinese government's policy of prohibiting foreign participation is likely to be relaxed for value-added network (VAN) services. In November 1994, a Chinese official stated that China would allow foreign telecommunications operators to enter the VAN business when China becomes a member of the WTO. According to a news report, foreign investors or firms will not be permitted to own telecommunications infrastructure, but they will be able to lease capacity from Chinese operators to provide services.

In early 1994, IBM was chosen to develop the Golden Bridge project—a national electronic data network—and work as a consultant for two other "Golden" projects: (1) Golden Customs, an electronic data interface for companies involved in foreign trade, and (2) Golden Card, a banking, credit card, and debit card transactions network.

Most observers believe that China will continue to welcome foreign investment in the telecommunications sector because, if it does not, it will not be able to meet its teledensity targets. Recent government statements are seen not as a barricade but rather as a shift toward a controlled step-by-step approach.

1997 Environment

Table A4.2 summarizes China's environment for telecommunications services in 1997.

Hong Kong

Highlights from offer at WTO Negotiations on Basic Telecommunications (WTO 1997):

■ Commits to international simple resale for facsimile and data transmission services.

■ Has already provided access to the local market for many basic telecommunications services, including voice and data transmission, mobile radio telephone, and mobile data services.

■ For local fixed-network services, has issued four licenses and will consider issuing further licenses in June 1998.

■ Commits to permit call-back and other alternative international calling services, certain satellite services, virtual private networks, and mobile satellite services.

■ Commits to the Reference Paper on regulatory principles.

Infrastructure and Policy

Telecommunications infrastructure in Hong Kong is one of the most advanced in the world not only in its teledensity but also the quality of the network: fixed line is a fully digitized fiber-optic network. On average, people in Hong Kong use telephone or other telecommunications services five times more than people in the United States. Moreover, in January 1997, Hong Kong became the first country to offer full number portability—that is, all fixed telephone users can keep their numbers when they move their home or office within Hong Kong or when they switch their subscription from one company to another. Wireless communication services, cellular phone, and pagers are also in wide use in Hong Kong. Almost 20 percent of residents had a cellular phone in 1997. Table A5.1 shows the growth of telephone services since 1986.

Hong Kong has a two-tier regulation system. The Telecommunications Authority (TA), a statutory body, oversees the regulation of the telecommunications sector; its executive arm is the Office of the Telecommunications Authority (OFTA), established in 1993. The TA and OFTA regulate and license telecommunications services, manage the radio frequency spectrum, and ensure the effective operation and successful development of Hong Kong's telecommunications industry. As part of its effort to attract foreign businesses and become a world financial center, Hong Kong is developing an advanced telecommunications network. As a result, it is also becoming an information hub for East Asia.

Table A5.1 Telephone subscribers, Hong Kong, 1986-96

	Main line	Cellular	Pager
1986	1,844,403	10,000	287,356
1987	1,988,524	28,060	376,843
1988	2,153,776	51,280	486,640
1989	2,304,572	89,193	594,216
1990	2,446,989	133,912	711,420
1991	2,596,397	189,664	880,666
1992	2,777,732	233,324	1,046,384
1993	2,955,563	290,843	1,244,392
1994	3,113,560	431,775	1,356,177
1995	3,254,349	687,600	1,323,741
1996	3,401,928	1,210,680	1,089,904

Source: Office of the Telecommunications Authority (OFTA). 1997. *Key Statistics for Telecommunications in Hong Kong*. <http://www.ofta.gov.hk> (17 July 1997).

Local and Long Distance Services

In 1994, HK Telecom's monopoly was broken, and competition was introduced into Hong Kong's local phone market. Three additional fixed-line operators—Hutchison Whampoa, Wharf Holdings, and New World Telephone—were invited to establish systems. All three are conglomerates, better known for their involvement in property than in telecommunications.

Hutchison Whampoa, a wireless communications operator, is a part of Li Ka-Shing's business empire; it is assisted by Australia's Telstra. Wharf Holdings has teamed up with Nynex to create New T&T Hong Kong, which provides voice, data, and video services. New World Telephone is a consortium led by New World Development (66.5 percent stake), US West (25 percent), Shanghai Long Distance (5 percent), and Infa Telecom Asia (3.5 percent). (Foreign telecommunications firms are not permitted to operate directly in Hong Kong, but they can provide technical assistance.) All three companies were founded with a large investment. New World Telephone, for example, was launched with an initial investment of HK$2 billion.

The Chinese government was supposed to be consulted on all Hong Kong infrastructure projects extending beyond 1997 (Lee 1994). Presumably, the three conglomerates consulted quietly with Beijing, because it has voiced no objection to the new licensing arrangement.

In international voice service, HK Telecom—of which Cable & Wireless of the United Kingdom owns 57.5 percent and CITIC owns 8 percent—still has a monopoly right.[5] However, this monopoly right does not cover a range of new services, and an emerging call-back

---

5. As of 1993, international telephone services accounted for 60 percent of HK Telecom's gross revenue; China services alone account for 26 percent (Lee 1993).

system is threatening its position. In the WTO Telecommunications Pact of February 1997, the Hong Kong authorities promised to permit the resale of international fax and data service as well as call-back services (because call-back was already legal, this promise amounted to an international binding of existing law). However, the benefits from the WTO Telecommunications Pact for consumers are expected to be relatively small because international calls in Hong Kong are already at rock-bottom prices: the cost of a call to the United States or Canada can be as low as 23 cents per minute using the call-back system. In 1996, in response to the widespread use of call-back systems, HK Telecom dropped its scheduled prices for international calls by about 50 percent. With lower revenues from international calls, HK Telecom may have to replace its flat fee for local service with a usage-based fee.

Mobile Communications Services

Hong Kong has the world's most liberal market for mobile communications. No other country has so many operators in such a small area and under so many different systems: seven cellular companies are operating under four different cellular standards: CT-0, CT-2, Personal Handy-Phone System (PHS), and Digital Enhanced Cordless Telecommunications (DECT). Another company joined the market in the summer of 1997. Furthermore, 34 licenses have been issued for paging operations.

The intense competition has resulted in low prices for cellular services: cellular users may pay as little as HK$1 (US$0.13) for a minute's air time, and a handset costs between a few hundred Hong Kong dollars (subsidized) and a few thousand (US$25 to US$250). The low prices and enhanced services offered by the competing firms have, in turn, stimulated demand for cellular phones. Cellular phone operators expect that by the year 2000, the number of cellular users will reach two million, almost 30 percent of Hong Kong's total population.

Other Services

No licensing restrictions apply to computer services and value-added services. International Value-Added Network Services (IVANS) were considerably liberalized when Hong Kong removed a requirement for bilateral government-to-government agreements to be in place prior to service rollout. The scope of IVANS was also widened to include value-added voice services, such as centralized answering and call-forwarding systems.

In May 1997, the TA licensed the 4 local phone operators and 14 new operators to administer Virtual Private Networks (VPNs). VPN is a service used mainly by multinational corporations for voice and non-voice communications with their overseas branch offices or affiliated

Table A5.2 Telecommunications services environment, Hong Kong, 1997

Type of service	Price (HK$)
Monthly charge, residential fixed line	67
Installation fee, residential fixed line	530
Local calls	free
IDD[a] call to New York per minute	2.98
Cellular monthly charge, standard plan	400
Installation wait for residential line	7 days

a. IDD (International Direct Dialing) is the cost of a phone call to New York using an incumbent carrier, HK Telecom.

Source: *Asian Wall Street Journal*, 10 June 1997.

companies. The authority announced that it would issue more licenses as applications are received and processed.

1997 Environment

Table A5.2 summarizes Hong Kong's environment for telecommunications services in 1997.

Indonesia

Highlights from offer at WTO Negotiations on Basic Telecommunications (WTO 1997):

- Deletes an economic needs test for new entrants in domestic mobile cellular telephone services, personal mobile cellular communication services, and regional and national paging services.

- Public voice telephony, circuit switched public data network, and teleconferencing services are currently supplied by a number of suppliers with exclusive rights; commits to a policy review to determine whether to admit additional suppliers upon the expiration of the exclusive rights (in 2011 for local service, 2006 for long distance service, and 2005 for international service).

- Offers competition for packed-switched public data network services, telex, telegraph, and internet access services, subject to use of networks of PT Indosat and PT Satelindo for international traffic.

**Table A6.1 Teledensity: number of subscribers,
Indonesia, 1990-95** (per 100 people)

Service	1990	1991	1992	1993	1994	1995
Fixed line	0.59	0.71	0.90	0.99	1.29	1.70
Cellular	0.01	0.01	0.02	0.03	0.04	0.11

Source: International Telecommunication Union, 1995 World
Telecommunication Report.

- Offers competition in domestic mobile cellular telephone services, paging, and public pay phone services.

- Limits foreign equity to 35 percent for all services except personal communication services, for which it requires joint ownership by a state-owned company.

- Commits to the Reference Paper on regulatory principles.

Infrastructure and Policy

By the end of 1995, Indonesia had 3.3 million installed main lines. Although the telecommunications infrastructure has developed rapidly
since 1985, when the teledensity was 0.4 lines per 100 people, the current teledensity of 1.7 lines is still one of the lowest among the APEC
countries. Table A6.1 shows the growth in teledensity from 1990 to 1995.

The policy planning and regulatory authority, the Ministry of Tour-
ism, Post, and Telecommunications (MTPT), exerts a strong voice in the
nation's telecommunications development. In August 1995, an MTPT decree
granted PT Telkom—a state-owned enterprise—the exclusive right to
provide domestic local service until 2001 and long distance services until
2006. For international service, PT Indosat—established by ITT of the
United States in 1967 and transformed to a 100-percent government-
owned corporation in 1981—was granted the exclusive right until 2005.

The national development plans, known as Repelita, have introduced
few structural changes in the telecommunications industry. The government's basic strategy is to attract international capital and foreign
advanced technology. In 1993, MTPT Decree 39 opened the Indonesian
telecommunications sector to foreign investors for the first time. Under
the five-year national development plan that ran from 1990 to 1994 (known
as Repelita V), PT Telkom constructed 2.2 million lines with technical
assistance from AT&T and from NEC of Japan. The next plan, Repelita VI
(1995-99), calls for US$12 billion in investment and an additional five
million main lines. To implement the plan, the government set up a Joint
Operation Scheme (known as JOS or KSO), which establishes a series of
consortia of domestic and foreign investors. Working closely with PT
Telkom, the consortia will develop regional telecommunications infra-

Table A6.2 Joint operation scheme, Indonesia, 1996

Region	Lines to be installed	Consortium member and shares (percentages)	
West Java	500,000	Domestic	52.5
		US West	35.0
		Asian Infrastructure Fund	12.5
Central Java	440,000	Domestic	50.0
		Telstra	20.0
		NTT	15.0
		others	15.0
Kalimantan	237,000	Domestic	53.2
		Telekom Malaysia	25.0
		others	22.8
Sumatra	460,000	Domestic	59.5
		France Telecom	35.0
		others	4.5
Eastern Indonesia	403,000	Domestic	55.0
		Singapore Telekom	45.0

Source: National Trade Data Bank, US Department of Commerce (1997).

structure and manage it for 15 years. The consortia will also share revenue with PT Telkom, which will take over the infrastructure after the contract period. According to the government blueprint, two million basic telephone lines will be installed by the consortia and an additional three million lines by PT Telkom. Table A6.2 gives details of the JOS plans.

Since Indonesia has more than 17,000 islands and extends about 5,000 km from west to east, laying a fixed-wire network would be prohibitively expensive. In 1976, therefore, Indonesia became the first developing country to own and operate a domestic satellite system, when the Palapa system (run by PT Telkom) was launched to provide television and telephone service to the Indonesian archipelago. The Palapa-B2R system (part of the Palapa series) improved transponder capacity to reach Thailand, Malaysia, Singapore, Philippines, and Papua New Guinea. In 1991, PT Telkom and PT Indosat set up a private corporation called Palapa Pacific Nusantra, which provides services to a wider region that includes Pacific islands and Hawaii.

Local and Long Distance Services

PT Telkom and PT Indosat still dominate the basic service markets; however, private companies, including foreign firms, have been emerging through a variety of joint ventures. At the same time, privatization of state-owned firms has begun.

In 1993, PT Telkom, PT Indosat, and PT Bimagaraha Telekomindo (a domestic private company) established PT Satellite Palapa Indonesia (PT Satelindo). PT Satelindo competes with PT Indosat in international phone services and also provides GSM cellular services using the nation's third satellite, Palapa C. In 1995, 70 percent of PT Satelindo's shares were offered to the public: 45 percent for domestic investors and 25 percent for foreign investors. PT Telkom and PT Indosat retained the remaining shares: 22.5 percent and 7.5 percent, respectively. DeTe Mobile, a subsidiary of Deutsche Telekom of Germany, bid successfully and became the only foreign owner.

Privatization of the state-owned operators, PT Indosat and PT Telkom, has also been under way. In October 1994, the government floated 25 percent of PT Indosat's shares on the New York Stock Exchange and 10 percent on the Jakarta Stock Exchange. This process raised US$1.16 billion in new capital. After this success, the government also tried to list PT Telkom's stock on the Jakarta and London Stock Exchanges. Expecting to raise around US$3 billion, the government initially planned to sell 27.5 percent of the shares. However, because the stock markets were low at that time, the government reduced the size of its initial offering to 19 percent and raised US$1.59 billion. About one-third of the 19 percent (6.5 percent of the company's capital) was sold internationally, while the rest (12.5 percent of the capital) was sold domestically. In December 1996, an additional 4.15 percent (388 million ordinary shares) was sold for US$0.6 billion. As a result, about one-quarter of PT Telkom is now owned by private investors.

Mobile Communications Services

Indonesia initially adopted two systems for its mobile network: the Nordic Mobile Telephone (NMT) system and the Advanced Mobile Phone System (AMPS). The Global System for Mobile Telecommunications (GSM) was added later, and the government plans to introduce a fourth system.

In 1986, PT Rahasa Hazanaha Perkasa (RHP) and PT Telkom entered into a revenue-sharing scheme to operate NMT-450. Later they jointly established Mobilsel, of which RHP owns 70 percent while Telkom and its pension fund the rest. Mobilsel is the solo NMT service provider and serves only about 30,000 subscribers. Nevertheless, it has been licensed to extend its operation over the entire nation.

The AMPS mobile phone networks is managed by three operators and serves more than 80,000 subscribers. PT Telkom owns a part of two joint-venture companies, PT Komselindo and PT Metrosel. PT Telekomindo Prima Bhakti, the third operator, is managed under a revenue-sharing scheme with PT Telkom, which owns 20 percent of the company.

Table A6.3   Telecom services environment, Indonesia, 1997

Type of service	Price (US$)
Monthly charge, residential fixed line	8.54
Installation fee, residential fixed line	241.80
Local call per 3 minutes	0.05
IDD[a] call to New York per minute	1.76
Cellular monthly charge, standard plan	20.49
Installation wait for a residential line	10 days

a. IDD (International Direct Dial) is the cost of a phone call to New York using an incumbent carrier, PT Telekom.

Source: Asian Wall Street Journal, 10 June 1997.

The GSM system, selected as a standard for mobile communications services, has 160,000 subscribers. Three operators are competing in the GSM system. The first operator is PT Satelindo. The second, PT Telkomsel, was jointly established by the two domestic monopolies, PT Telkom and PT Indosat; 17 percent of its shares were then sold to PTT Telecom, a Dutch telecommunications operator, for US$304 million. An additional 22 percent of Telkomsel is jointly owned by Nynex of the United States, Mitsui of Japan, and the Asian Infrastructure. The third operator, PT Excelcomindo, is more domestically oriented: the domestic AMPS operator, Telekomindo, owns 60 percent of its shares.

Despite these varieties of systems and multiple operators, the mobile communications market in Indonesia is hardly competitive. PT Telkom's monopoly power is protected by tight government regulations, revenue-sharing schemes, or the capital ownership structure of the mobile operators.

Besides providing telecommunications to a population dispersed over many islands, radio communications systems can be used to create fixed-wireless networks using radio-based stations in urban areas. Fixed-wireless networks can be installed more quickly and cheaply than ordinary telephone lines, although mobile phone services cannot be operated with these networks. In 1994, the government gave a license for a fixed-wireless operation to Ratelindo, a joint venture between PT Telkom and Bakrie Electronics, a private company. Ratelindo was licensed to provide 280,000 fixed-wireless connections (250,000 in Jakarta and 30,000 elsewhere in Java). Ratelindo's fixed-wireless system is assisted by Hughes of the United States.

1997 Environment

Table A6.3 summarizes Indonesia's environment for telecommunications services in 1997.

Japan

Highlights from offer at WTO Negotiations on Basic Telecommunications (WTO 1997):

- Deletes the reservation concerning international simple resale of voice services.

- In April 1996, agreed to remove long-standing foreign equity limits on Type I carriers and radio-based services, leaving only two companies, KDD and NTT, with foreign equity limits (at 20 percent). Aside from these company-specific restrictions, commits to open market access in all market segments for basic telecommunications services (facilities-based and resale).

- Commits to the Reference Paper on regulatory principles.

Infrastructure and Policy

Japan's highly developed telecommunications infrastructure is represented by the number of installed telephone lines—more than 60 million, or about 48 lines per 100 persons. According to a survey by the Ministry of Post and Telecommunications (MPT), Japan's telecommunications regulatory and policy authority, in 1995 the total equipment investment of the telecommunications industry was 3.5 trillion yen (about US$30 million), an increase of 22 percent from 1994. The Telecommunications Council, an advisory body of MPT, has estimated that the output share of the telecommunications sector in Japan (including telecommunications equipment) will exceed 8 percent in 2000, far more than the 4-percent share of the automotive and transportation sector. The telecommunications sector will also create an additional 500,000 jobs and employ 1.5 million workers by 2000, up from 1 million in 1994.

Japan is now constructing an advanced telecommunications infrastructure corresponding to dramatic technological progress in the sector. In a 1994 report, the Telecommunications Council proposed installing a nationwide fiber-optic network reaching individual subscribers. This network would provide highly integrated multimedia communications, including voice, text, image, and motion video data. According to the report, the fiber-optic network should reach 20 percent of the population by 2000, and 100 percent by 2010.[6] Recently, the Ministry of Construction announced that by 2010, it will construct 150,000 km of

---

6. As of 1994, only 6.6 percent of the subscribers' network consisted of fiber-optic cable, whereas 96 percent of the trunk network was fiber-optic.

underground tube to carry fiber-optic line. These tubes will be located along main traffic roads throughout the nation and will be available for lease by any telecommunications carrier for a low fixed charge.

The three large actors in Japan's telecom industry have long been MPT; the government-owned Nippon Telegraph and Telephone Public Cooperative (currently known as Nippon Telegraph and Telephone Company, or NTT), a virtual monopoly in the domestic market; and the government-owned Kokusai Denshin Denwa Cooperation (KDD), the dominant international carrier. This picture has changed gradually since a major deregulation process was initiated in 1984. To stimulate market competition, many private companies have been allowed to enter the basic telecommunications services market. The carriers are divided into two groups: "Type I" carriers own and operate basic networks, and "Type II" carriers provide value-added services using a network leased from Type I carriers. Japanese law permits up to 33 percent foreign participation in Type I carriers and imposes no limit on foreign participation in Type II carriers. As of January 1996, there were 123 Type I carriers and about 2,805 Type II carriers.

As other carriers entered the picture, the two giant telecommunications monopolies began competing with each other by offering comprehensive services: NTT now offers international service, and KDD offers domestic service. A further step will be taken in 1999, when the giant telecommunications carrier NTT will be broken up into one long distance company and two local carriers.

In February 1996, MPT announced a deregulation package for the "Second Reform of the Info-Communications System in Japan," aimed at stimulating rate reductions by decreasing the MPT's role in price regulation. According to the plan, since March 1997, MPT requires mobile carriers only to give prior notification of rate changes, instead of having to seek prior authorization.

In the cable TV business, some deregulation has taken place. In 1994, MPT lifted the regulations on cable TV operators; now a single company may own more than one local cable station and thus may offer a nationwide network. The government also permits up to 33.3 percent foreign ownership of cable TV firms.

Local and Long Distance Services

In 1986, the government began to privatize NTT, the largest company in Japan. In the world's largest privatization, three blocks of NTT shares were sold for US$12.4 billion. A fourth offer of NTT shares was shelved because of the depressed stock market. In 1997, one-third of NTT was privately owned, while the rest was held by the government. The government intends ultimately to reduce its ownership to one-third of total shares.

Table A7.1 Cost of domestic long distance call between Tokyo and Osaka,[a] Japan (yen)

	NTT	NCCs	Other
1985	400	—	—
1993	180	170	—
1996	140	130	110

NCC = New Common Carriers.

a. Cost is for 3 minutes during the day.

Source: Ministry of Post and Telecommunications (Japan) (1997).

In spite of the privatization, NTT's activities are strictly regulated by the law; the company cannot even appoint directors without MPT approval. It remains to be seen whether tight government control will continue after the government becomes a minority shareholder.

Initially, foreigners were not permitted to purchase NTT or KDD shares. The government lifted these prohibitions in 1992, and foreigners can now buy up to 20 percent of the shares in either company.[7] However, the high price of NTT and KDD shares and the fact that foreign corporate owners are unlikely to have an influential voice in management have discouraged foreign carriers from buying significant blocks of NTT equity.

Beginning in 1987, NTT's monopoly in the domestic long distance service market was challenged by three new entrants: Daini-Denden (DDI), formed mainly by Kyocera, Sony, and Mitsubishi; Japan Telecom, formed by state-operated railway companies (JR East, JR West, and JR Tokai); and Teleway Japan, in a consortium with Toyota. No foreign telecommunications firms participated in these consortia. As of September 1995, the new entrants collectively had 31 million subscribers, while NTT had 61 million. Once competition was introduced, the price for a long distance call between Tokyo and Osaka (daytime, three minutes) fell by more than 65 percent in 10 years (table A7.1).

In the local service market, the Tokyo Telecommunications Network (TTNet), a subsidiary of the giant Tokyo Electric Power Co., (Tepco), established a 20,000-km fiber-optic network in the Tokyo area and started using this network to deliver local phone service. TTNet plans to expand the network nationwide and enter the mobile phone and cable TV

---

7. NTT shares were listed on the New York and London Stock Exchanges in late 1994.

Table A7.2   Cost of an international call between Japan and the United States,[a] 1985 and 1995 (yen)

	NTT	NCCs
1985	1,350	—
1995	480	470

NCC = New Common Carriers.

a. Cost is for 3 minutes during the day.

Source: Ministry of Post and Telecommunications (Japan) (1997).

markets.[8] Since November 1994, cable TV companies have also been able to apply for telephone licenses. However, NTT retains its monopolistic status in the local market, providing for 99 percent of the domestic market. In fact, in 1994, nearly 50 percent of NTT's revenue, 3.1 trillion yen, came from the interconnection service fees paid by the three largest new entrants. The next plan was to introduce effective competition into the local phone market after 1996 by creating a fair and objective rule governing interconnections between NTT and other local or long distance carriers.[9]

For international service, KDD has faced competition since 1989: its competitors include International Telecom Japan (ITJ), formed mainly by a Japanese business group—Mitsubishi and Mitsui; and International Digital Communications (IDC), formed by a group including Itochu, Toyota, Cable & Wireless (United Kingdom), and Pacific Telesis (United States). Both new entrants are offering international calls at lower rates than KDD. The cost of international calls is now almost 65 percent lower than at the beginning of liberalization (table A7.2). Between 1984 and

---

8. Six companies, all affiliates of local electricity firms like TTNet, are already licensed to run local networks. At this stage, the margins from the local business are so thin that only TTNet has good prospects. In fact, even NTT's local business is unprofitable, since for nearly two decades the rate for a local call has been frozen by the Ministry at 10 yen (about 10 cents) for three minutes.

9. In August 1994, a group of 26 Type II telecom carriers, mostly value-added network operators, submitted a report to the MPT Minister, claiming that NTT's interconnection charges were so high that about 600 out of the 1500 Type II operators were losing money. According to the report, interconnection charges in Japan are several times higher than those of other industrial countries. The report also claimed that NTT discriminated in favor of its affiliated operators, intentionally delaying interconnection with new entrants that plan to start new services (such as virtual private network services). The report demanded that the MPT put an end to such price discrimination and create an environment in which nonaffiliated operators can compete fairly with NTT's affiliated operators. In December 1994, MPT finally issued an interconnection order to NTT.

Table A7.3   Increase in international telephone traffic in selected countries, 1984 and 1994

	Millions of minutes		Rate of change (percentages)
	1984	1994	
Australia	160	645	304
Canada	159	787	395
France	1,008	2,500	148
Germany	1,662	4,960	198
Hong Kong	123	1,578	1182
Italy	503	1,760	250
Japan	191	1,411	639
Singapore	95	560	489
Taiwan	46	502	996
United Kingdom	1,122	3,221	187
United States	1,883	13,121	597
Average			347

Source: International Telecommunication Union, 1995 World Telecommunication Development Report.

1994, the volume of international calls increased about 639 percent, significantly more than the average increase in other high-income countries (table A7.3).

Despite these rate reductions, Japan's international telephone rates still remain among the highest in the world. Several US carriers are offering call-back services to provide a cheaper alternative to Japanese customers.[10] A mid-size Japanese company could save about 40 percent.[11] KDD and two other Japanese companies petitioned the government to investigate call-back services.

Mobile Communications Services

In 1979, Japan became one of the first countries to introduce a mobile cellular phone system when NTT commercialized a car phone service. The commercialization of the new technology began slowly, but it has started to increase. In 1994, there were only 4 million mobile phone subscribers (about 3.5 percent of the population), but by July 1996, there

10. For a description of call-back service, see the Overview section. As of September 1994, about 10 companies offered this service in Japan.

11. Savings are calculated on the basis of a initial monthly international phone bill of US$5,000 to Europe and North America.

were 16.9 million. This increase is due mainly to the emergence of a new mobile phone service, Personal Handy-phone System (PHS).

NTT started its cellular operation in 1979 as a sole provider. In 1985, two additional providers joined: Japan Cellular, a subsidiary of DDI, which started its operation in Osaka; and IDO, formed by Telway Japan and Toyota, which focuses on Nagoya and Tokyo.

DDI was the only company to adopt Motorola's Total Access Communication System (TACS). The Japanese authorities restricted Motorola's activities by excluding the TACS system from the densely populated Tokyo-Nagoya region. As a consequence of US pressure, IDO was assigned responsibility for building up the TACS system in 1990. The problem was that IDO had already committed considerable sums to building an infrastructure based on NTT's proprietary technology (called NTT Mobile Telephone System)[12] and therefore had little incentive to invest further in the TACS system. As a result, by early 1994, while the TACS system had taken a 50-percent share in the Osaka area, it claimed less than 2 percent in the Tokyo-Nagoya market. The political pressure arising from the Motorola dispute in March 1994 forced IDO to expand its TACS infrastructure. By the end of 1994, the TACS system covered 94 percent of the population, up from 61 percent at the end of March 1994. IDO's monthly market share (including both NTT and TACS systems) rose from 6 percent in March to 45 percent in December 1994, reflecting Motorola's very competitive cellular phone rates.[13]

Two recent entrants into the mobile communications market—the Digital Phone group, formed by Japan Telecom, and the Tu-ka Cellular group, owned by a consortium under Nissan's lead—operate nationwide.

The government took the initiative in introducing PHS as a cheaper and fully digitized mobile service. NTT Personal Communications Network group (NTT Personal, mostly owned by NTT and NTT Mobile Communications Network) and DDI Pocket Telephone group (DDI Pocket, a subsidiary of DDI) started their operation in July 1995. In October 1995, the Astel group started its operation.[14]

Because PHS uses a weaker signal than conventional cellular, people cannot use it in fast-moving vehicles, such as cars and trains. However, the relatively simple technology of PHS reduces the fixed and operating costs as well as the price; for example, with PHS a three-minute call costs only 40 yen, whereas it costs about 200 yen with a conventional

---

12. For cellular standards, see the Overview section.

13. At the end of 1994, the respective market shares were 57 percent for the NTT analog system, 28 percent for the Motorola analog system (TACS), and 14 percent for new digital systems (*Nihon Keizai Shimbun*, 10 January 1995).

14. The Astel group is owned by various trading and telecommunications companies, including KDD, TTNet, Japan Telecom, and Teleway Japan.

Table A7.4a Mobile phone service subscriber: conventional cellular phone subscriber, Japan, 1990-95

	Number of Subscribers	Rate of change (percentages)
1990	868,078	
1991	1,378,108	58.8
1992	1,712,545	24.3
1993	2,131,367	24.5
1994	4,331,369	103.2
1995	8,051,892	85.9

Table A7.4b Mobile phone service subscriber: conventional cellular service versus PHS, Japan, 1995-96

	Conventional	PHS
Number of subscriber as of 1995/6	8,051,892	81,203
Number of subscriber as of 1996/7	13,650,334	3,230,167
Increase	7,830,092	3,148,964

Source: Ministry of Post and Telecommunications (Japan) (1997).

cellular service. PHS also has an advantage in terms of data transmission speed, since it is fully digitized. The PHS carriers plan to provide high-speed data transmission service at 32K BPS in 1997, while the conventional cellular phone services offer 9.6K BPS.

The number of PHS subscribers is not growing as fast as was expected (tables A7.4a and A7.4b). However, the cellular service operators cut their prices to compete with PHS: the basic monthly rate was reduced from 30,000 yen in 1985 to 7,400 yen in 1995, and the installation fee from 80,000 yen to 6,000 yen over the same period. The cellular operators also aggressively advertised their advantage, a wider service area. As a result, the conventional cellular market has expanded very rapidly since the introduction of PHS—a good example of the benefits of introducing competition.

Other Services

The cable TV business is getting more attention in Japan. For a long time, cable TV was considered to be supplemental to traditional TV service, especially in areas that had poor TV reception. Often, local gov-

Table A7.5   Telecom services environment, Japan, 1997

Type of service	Price (US$)
Monthly charge, residential fixed line	13.80
Installation fee, residential fixed line	567.60
Local call per 3 minutes	0.08
IDD[a] call to New York per minute	1.30
Cellular monthly charge, standard plan	30.74
Installation wait for residential line	2.5 days

a. IDD (International Direct Dial) is the cost of a phone call using an incumbent carrier, KDD.

Source: Asian Wall Street Journal, 10 June 1997.

ernments own all or part of the local cable operators in their communities; under such circumstances, the cable service is likely to offer only the retransmission of terrestrial and satellite TV programs. By contrast, relatively large operators that offer more than five cable programs are distinguished as "urban CATV" operators, and they are more commercially oriented. As of 1994, 162 urban CATV operators had 2.2 million subscribers out of 10.4 million total cable subscribers.

Competition in the multichannel broadcasting market is expected to become stiffer. Two satellite digital broadcasting services, each offering 50 to 100 channels, were operating at the end of 1996: DMC, established by Japan Satellite Systems (a major Japanese satellite company, operating JSAT satellites) and four major Japanese trading companies; and DirecTV Japan, established by Hughes Network Group and Mitsubishi (which also owns part of another major Japanese satellite company, Space Communications). Recently, an Australian media giant, News Corporation (J-Sky-B), also announced that by 1998 it would start a satellite broadcasting service with 150 channels.

Faced with this competition, the urban CATV companies are seeking new service opportunities. Some of them are planning to provide cable telephony service, competing with NTT. But most of the companies are seeking value-added services, such as data transmission or interactive TV, to make best use of their high-capacity cable networks. The larger operators began to offer internet access in 1997.

1997 Environment

Table A7.5 summarizes Japan's environment for telecommunications services in 1997.

Table A8.1 Teledensity and cellular subscription, South Korea, 1990-95

	1990	1991	1992	1993	1994	1995
Teledensity,[a] fixed line	30.96909	33.67826	35.71565	37.87122	39.69992	41.47203
Cellular subscription[b]	0.19	0.38	0.62	1.07	2.16	3.66

a. Teledensity is the number of main line per 100 people.
b. Percentage of population who subscribe to cellular phone service.

Source: International Telecommunication Union, 1995 World Telecommunication Development.

South Korea

Highlights from offer at WTO Negotiations on Basic Telecommunications (WTO 1997):

■ Increases foreign equity participation limit on facilities-based suppliers from 33 percent to 49 percent after 2001.

■ Increases limit on foreign equity in the national supplier, Korea Telecom, from 20 percent to 33 percent after 2001.

■ Permits competition in wire-based telephone services never before opened to full competition.

■ Permits full competition in supply by resale of all telecommunications services except voice, without phase-in; phases in international simple voice resale by 2001.

■ Permits market access for domestic voice resale as of 1999, when it will allow foreign equity participation up to 49 percent, rising to 100 percent by 2001.

■ Commits to the Reference Paper on regulatory principles.

Infrastructure and Policy

Telecommunications services in South Korea were nurtured by its fast economic growth, and the government's focus on telecommunications infrastructure. In 1994, the number of installed main lines reached 16 million, or about 36 lines per 100 people—a significant increase from 1982, when there were under 3 million lines installed (table A8.1).

South Korea is one of only 10 countries that develop and export electronic switching systems. Since 1986, the nation's network has installed and exported the locally designed digital exchange. Because the government promotes exports, South Korea's export of telecommunications equipment to the United States amounted to US$593 million in 1995. There has been strong international pressure to open South Korea's telecommunications equipment and service as well as its service market. The Information Technology Agreement and the WTO Basic Telecommunications Agreement are expected to open the door for both markets.

Until the South Korean government issued deregulation plans in 1990, monopoly operators were designated to provide specific services: Korea Telecom (KT), the national fixed-line service operator; Korea Mobile Telecom (KMT), a cellular phone operator started in 1984; and Dacom, a data communication service provider. All three operators were government-owned companies until recently; they are now being privatized. The government originally intended to sell 49 percent of KT's shares, more than 50 percent of KMT, and 100 percent of Dacom. It was successful in selling a majority of KMT's shares in 1989. However, it has been able to sell only a small percentage of KT's and Dacom's shares.

Apparently, domestic investors failed to buy the shares because the Ministry of Communications (MOC) was unwilling to disclose information about the companies' affairs. Foreign investors were excluded from the bidding because MOC considers both companies to be strategic national industries. Furthermore, the MOC prohibits telecommunications equipment manufacturers from owning more than 8.3 percent of the shares, so Goldstar Telecommunications, Hyundai Electronics, and Samsung Electronics were also excluded from bidding since each of them already held the maximum share of Dacom. In addition, MOC rules prohibit any single individual or corporation from owning more than a one-third share. Because of all these restrictions, investors were not sure whether the MOC would allow private firms to exercise managerial control over Dacom, or would let KT alone. These doubts were reinforced by the fact that almost all of Dacom's top managers were former ministry officials or KT employees.

In 1994, the MOC announced another set of deregulation plans. The classification of telecommunications service providers was simplified, so that a single telecommunications company is now able to provide a wider range of services. In March 1995, monopoly in the long distance market ended when Dacom was allowed to enter.

In 1995, the MOC's successor agency, the Ministry of Information and Communication (MIC), proposed further deregulation. According to the 1995 plan, new licenses would be issued for almost all the major telecommunications markets except for the local call market, which is dominated by KT. Thus, seven different markets were subject to liberalization: international call, PCS, trunked radio service, CT-2, leased lines, paging, and wireless data communications. Some 30 licenses were issued for these various services in June 1996.

Even after the steps toward deregulation, the MIC retains its regulatory role over the telecommunications market. Market entrance and lines of business remain subject to the MIC's approval. The MIC plans to introduce more competition and give somewhat greater scope to foreign carriers.

Local and Long Distance Services

Under the MOC, KT operated both domestic and international networks as a state-owned monopoly. At one time, KT held a 33-percent share in Dacom; this share was sold in 1990 so that the two companies could compete in each other's turf. Although KT continues to dominate South Korea's basic telecommunications services, a certain degree of competition has been introduced in long distance and international services. For instance, Dacom now offers international service at prices that are 3 to 5 percent lower than KT's. A significant number of customers have switched to Dacom, although initially Dacom's service was limited to the United States, Japan, and Hong Kong.[15] Dacom's attractive price regime is determined not by the market but by the MIC, which will equalize the price of international calls once Dacom captures 30 to 40 percent of the market. Thus, real price competition in the international market seems a distant goal.

Since March 1995, Dacom has also offered domestic long distance service. KT retains its monopoly only in the local call market, which will open to competition by 1999.

Mobile Communications Services

In recent years, Mobile communications service has attracted considerable attention since it has had the most prominent growth opportunity. KMT started cellular phone and paging services in 1984. As of 1993, it had only a half-million subscribers; however, the market is expected to expand to more than four million subscribers by 2000.

Following the deregulation plan of 1990, 10 new paging operators were licensed for nine regional markets, including two in Seoul. In 1992, a new cellular license was awarded to Sunkyong. But because of a political scandal in 1992,[16] the final decision was suspended for another

---

15. Dacom's service was expanded to 65 countries in 1993.

16. In 1992, the presidential candidate Kim Young Sam pressured a consortium led by the Sunkyong group, the nation's fifth-largest business group, to surrender the license it had just been granted by outgoing President Roh Tae Woo. Three foreign firms—GTE of the United States, Hutchison Communications of Hong Kong, and Vodafone of the United

two years. In January 1994, Sunkyong pulled out of the proposed cellular phone consortium and instead acquired from KT a 23-percent stake in KMT. In February 1994, the government granted the country's second cellular communications network license to another consortium, Sinsegi, led by South Korea's steel giant, Pohang Iron & Steel, with a 15-percent stake, and a textile company, Kolon Group, with a 14–percent stake. About 50 percent of the Sinsegi shares were distributed among several large South Korean business groups, and about 22 percent were reserved for four foreign participants: Pactel, Southwestern Bell, GTE, and Qualcomm.

The US companies were chosen because of their advanced cellular technology, the Code Division Multiple Access (CDMA) system. South Korea intends to acquire the CDMA technology and eventually export CDMA equipment. But it is still uncertain whether CDMA will become a world standard. In fact, South Korea will be the first country to adopt as standard the CDMA system rather than the TDMA system, which is widely used in industrial countries.

Preparing for competition with Sinsegi in the cellular phone market, KT reduced its installation fees by half and the monthly base charge by 20 percent. However, it increased the cost of a call by 30 percent.

PCS is growing popular and is expected to become the major mobile phone service in the next century. In June 1996, KT, LG Telecom, and Hansole PCS were awarded PCS licenses. The distribution of licenses was grounded in politics, not economics; the MIC decided that the three licenses should be granted to three different types of companies or groups: KT was a "major telecommunications provider," LG Telecom was a *chaebol*, and Hansole PCS was a group of small-medium companies. The MIC announced that PCS would adopt the CDMA system as its standard. LG Information & Communication, a telecommunications equipment manufacturer belonging to the LG group, is mainly developing CDMA.

Kingdom—were also part of the Sunkyong consortium. The reason given for revocation was that Sunkyong appeared to have won its license largely because Roh's daughter was married to the consortium's leader. Kim promised that after he became president, the license would be awarded to the most qualified group of bidders.

Initially, the Kim Young Sam government wanted to prevent the country's leading telecommunications companies—which belonged to South Korea's big business group—from dominating another sector of the economy. Despite their experience in telecommunications, these four companies were effectively barred from leadership roles by laws prohibiting them from owning more than 10 percent of any consortium. But President Kim finally asked the Federation of Korean Industries, an association of *chaebol*, to choose a single consortium. Ironically, the head of the federation was the leader of Sunkyong.

Table A8.2 Telecom services environment, South Korea, 1997

Type of service	Price (US$)
Monthly charge, residential fixed line	2.80
Installation fee, residential fixed line	280.02
Local calls per minute	0.05
IDD[a] call to New York per minute	1.18
Cellular monthly charge, standard plan	23.52
Installation wait for residential line	1/2 day

a. IDD (International Direct Dial) is the cost of a phone call to New York using an incumbent carrier, Korea Telecom.

Source: Asian Wall Street Journal, 10 June 1997.

Other Services

Since Dacom was established in the mid-1980s, it has dominated data communications and value-added services. However, the MIC has introduced competition in value-added services such as an Integrated Services Digital Network (ISDN), EDIN, and e-mail. Initially, foreign participation was limited in this area, but the restriction was lifted in 1994 as a result of US-South Korea telecommunications talks.

1997 Environment

Table A8.2 summarizes South Korea's environment for telecommunications services in 1997.

Mexico

Highlights from offer at WTO Negotiations on Basic Telecommunications (WTO 1997):

- Raises the foreign equity limitation from 40 percent to 49 percent for all telecommunications service suppliers and ends the exclusivity of regional duopolies in cellular telephony.

- Commits to competition in all market segments of public telecommunications services on facilities and resale bases: voice telephone service, data transmission, private-leased circuit services, paging, and certain cellular telephone services.

Table A9.1 Telephone subscribers, Mexico, 1988-95

	Main line	Cellular
1988	4,387,436	1,500
1989	4,847,166	8,500
1990	5,354,500	63,926
1991	6,024,800	160,898
1992	6,753,652	312,647
1993	7,620,880	386,132
1994	8,492,521	569,251
1995	8,801,030	642,000

Source: The International Telecommunication Union, 1995 World Telecommunication Development Report.

- For cellular telephony, allows more than 49 percent foreign investment, subject to prior authorization.
- Commits to the Reference Paper on regulatory principles.

Infrastructure and Policy

Mexico had installed 8.8 million main lines by 1995. Although the number of main lines had increased by more than 10 percent per year since 1990, the country had fewer than 10 lines per 100 people (see table A9.1).

Until 1990, Teléfonos de México (Telmex) dominated Mexico's basic telecommunications, with a 90-percent market share. The Secretariat of Communications and Transportation (SCT), Mexico's telecommunications regulatory authority, provided other services, such as telegraphy and satellite access.

In 1990, to reap the benefits of competition and new technology, the government decided to transfer the SCT's telegraphy and satellite communications services to a newly established public company, Telecommunications de México, and to privatize Telmex. The government's 56-percent stake, held in the form of AA shares, was converted to 20.4 percent of new AA shares, 5 percent of A shares, and 31 percent of L shares. Both AA and A shares have full voting rights, but only Mexican nationals could own AA shares. L shares had limited voting rights. The AA shares were sold to a consortium led by a Mexican holding company, Grupo Carso. Southwestern Bell and France Telecom were involved in this consortium. In 1997, Grupo Carso owned 10.5 percent of

Telmex's shares, Southwestern Bell owned 10 percent, and France Telecom held 5 percent. Meanwhile, most telecommunications services had been liberalized. Telmex had an exclusive right to provide domestic long distance and international services until the end of 1996. Beginning in 1997, Telmex opened its local network for interconnection.

Under the Federal Telecommunications Law of 1995, all telecommunications services—including local, long distance, international, and satellite—should be provided in a fully competitive manner. SCT was ordered to process applications for a Public Telecommunications Network (PTN) license immediately. In principle, a PTN operator can provide all telecommunications services. This means that a CATV operator, for example, could provide telephony services using a PTN license.

The 1995 law also specifies interconnection procedures: interconnection arrangements should be established within 60 days of a request, on a nondiscriminatory basis. Telmex was directed to finish rebalancing its long distance and local call service rates by 1997, when the long distance market was liberalized. Telmex will separate its financial accounts by type of service, and it is not supposed to subsidize a sector (e.g., local service) that faces competition. However, the law limits foreign ownership to 49 percent.

Local and Long Distance Services

Telmex has enjoyed monopoly status in the Mexican telecommunications market for some time and has operating margins above 40 percent. However, as it faced the prospect of competition in the profitable long distance market, Telmex began to improve its notoriously bad service. The company has invested more than US$11 billion since the early 1990s.

Although competitive access to the local market has in principle been possible since 1990, few firms have yet entered this area. This fact may be partly explained by the large cross-subsidies Telmex provides to local service from its long distance operations. Once cross-subsidies are eliminated, potential competitors to Telmex in the local call market will be wireless communications companies and CATV operators. Grupo Iusacell has been constructing a local network with wireless technology (Bell Atlantic of the United States owns 42 percent of the company). For CATV operators, Telmex has tried to take strategic control in this segment. In 1995, Telmex negotiated with Mexico's media giant, Televisa, to take a 49-percent stake of Cablevision, Televisa's cable services arm. This alliance is controversial from the standpoint of competition, since Televisa controls more than 80 percent of the TV broadcasting market and Cablevision is the monopoly in Mexico City.

Table A9.2 Outgoing international telephone traffic, Mexico, 1985-95

	International calls from Mexico (millions of minutes)
1985	144
1986	178
1987	197
1988	211
1989	315
1990	415
1991	501
1992	684
1993	760
1994	844
1995	945

Source: International Telecommunication Union, 1995 World Telecommunication Development Report.

More competition is expected in the US$4 billion Mexico-US long distance market. US telecommunications operators have been very aggressive in the Mexican international market because more than 80 percent of Mexico's international calls are to the United States, and ratification of NAFTA prompted the integration of North American telecommunications. In fact, the volume of Mexican outgoing calls has been steadily increasing during the past decade (see table A9.2)

MCI Communications formed a US$1 billion joint venture, Avantel, with Mexico's largest financial group, Grupo Financiero Banamex-Accibal (Banacci). MCI owns 55 percent and Banacci 45 percent of the venture. This alliance gives MCI access to Banacci's business network in Mexico. MCI already has a strategic alliance with the Canadian telecommunications giant Stentor; hence the Avantel initiative creates an end-to-end North American communications network.

In 1994, AT&T formed a US$1 billion telecommunications joint venture with the large industrial group Grupo Alfa. The venture, named Alestra, merged in 1996 with another telecommunications consortium, Unicom, led by GTE of the United States, Telefónica of Spain, and GF Bancomer, Mexico's second-largest bank group. The Mexican companies own 51 percent of the joint venture shares (25.6 percent by Alfa and 25.4 percent by Bancomer), and the internationals hold 49 percent (20 percent by AT&T and 14.5 percent each by GTE and Telefónica).

Besides these two joint ventures, five companies have been granted licenses to compete with Telmex in the long distance market: Cableados Y Sistemas, Investcom, Iusatel, Marcatel, and Miditel. Bell Atlantic participates in Iusatel, while Teleglobe and IXC of the United States take part in Marcatel. In 1995, before the Cableados Y Sistemas and Miditel licenses were granted, six potential Telmex challengers (Unicom and Alestra were then independent) formed the Association of Telecommunications Concessionaires (Actel) to carry out the tough interconnection negotiations with Telmex.

In the face of looming competition, Telmex had its own surprise: it announced an alliance with the US long distance telecommunications operator, Sprint, which is expected to cooperate with Grupo Iusacell.

Mobile Communications Services

Mexico's cellular concessions are divided into nine regions. For each region, two licenses were issued—one local and one national. Telmex holds the national concession, which it serves through its Telcel unit. The local cellular licenses were auctioned in 1989 and are now owned by Mexico's large business groups.

The Mexican government can grant concessions for fixed wireless phone services at any time, although, to protect Telmex's financial position, it has granted concessions only cautiously. Grupo Iusacell, Mexico's largest independent cellular phone company, already has a national fixed wireless concession with a focus on Mexico's large cities.

Alfonso Romo Garza, head of Grupo Pulsar, one of Mexico's large conglomerates, is also seeking a fixed wireless concession. He proposes to focus as much as 40 percent of his concession on rural areas, where it would be extremely costly for Telmex to install lines (30 percent of Mexicans live in rural areas). Romo Garza has acquired a 15-percent stake in the British fixed wireless telephone company Ionica, and is securing licensing rights for Mexico and other Latin American countries.

Nextel Communications, a New Jersey company that offers mobile radio services in major US cities, announced it would purchase a 22-percent stake in Mobilcom, a leading wireless telecommunications provider in Mexico, to create the first all-digital wireless communications network in North America. Nextel had earlier announced a similar deal with Clearnet Communications, a leading Canadian wireless communications company. Nextel and Mobilcom have agreed to share frequencies along the US-Mexican border. Licenses granted by the Mexican government to companies owned by Mobilcom cover Mexico City, Guadalajara, Monterrey, Acapulco, and Tijuana, as well as Mexico's major highways. Protexa, an industrial group that aims to be a comprehensive telecommunications services provider, has formed a joint venture with Motorola to provide cellular phone service linking sites in half of northern Mexico. Motorola had already acquired one of Mexico's cellular concessions.

New Zealand

Highlights from offer at WTO Negotiations on Basic Telecommunications (WTO 1997):

- Commits to open markets for all basic telecommunication services for all market segments (local, long distance, and international).

- Permits no single foreign entity to hold more than 49.9 percent of Telecom New Zealand, but does not limit the overall foreign shareholding in that operator.

- Commits to the Reference Paper on regulatory principles.

Infrastructure and Policy

In 1995, 1.72 million main telephone lines, or 48 lines per 100 inhabitants, served virtually every house and office in New Zealand. New Zealand has digitized 99 percent of the main line network—the highest percentage in the world. This advanced infrastructure is a result of the government's market-oriented policy.

Until the telecommunications reform started in 1987, the New Zealand Post Office was the only service provider. Following massive reforms in the early 1980s, the New Zealand Post Office was divided into three public companies to provide three services: postal, banking, and telecommunications. The telecommunications company was Telecom of New Zealand (Telecom NZ). At the same time, the responsibility for telecommunications policy was transferred from the Post Office to the Department of Trade and Industry (which became the Ministry of Commerce in 1988). At this stage, the telecommunications market was not fully deregulated, and only a few large companies with their own infrastructures—namely, railroad and electricity companies—had telecommunications networks for their internal use.

Once further deregulation took place in 1989, the government allowed full access to the telecommunications market, reflecting its belief that market mechanisms offer the best way to achieve efficient telecommunications. This mindset partly explains why New Zealand has no sector-specific regulator but simply relies on a general competition law in the Commerce Act of 1986.

In 1990, Telecom NZ was privatized. Bell Atlantic and Ameritech of the United States purchased the company for A$4.25 billion. As required by the contract, they sold more than half of the shares on the New Zealand stock markets by 1993, and each firm now owns 24.8 percent of Telecom NZ. Despite the government's procompetition stance, Telecom

Table A10.1 Price change in long distance call tariff, New Zealand, 1987 and 1995

Telecom's long distance call tariff	Peak time		Off-peak time	
	Shortest distance tariff	Longest distance tariff	Shortest distance tariff	Longest distance tariff
1987 (cents/minute)	6.50	144.00	2.20	47.00
1995 (cents/minute)	6.55	102.50	2.20	29.40
Change (percentages)	-0.76	-28.82	0.00	-37.45

Sources: Telecommunications Reform in New Zealand (1987-95); New Zealand Ministry of Commerce (1995).

NZ remains a monopoly in the local service market. Even after privatization, the government holds Telecom NZ's "Kiwi share," and it requires Telecom NZ to provide universal service and apply a price cap system on residential charges.

In the long distance and international services, however, deregulated privatization led to market competition. New entrants include Clear Communications, established by Bell Canada and MCI International of the United States (each of them has a 25-percent share), Television New Zealand, New Zealand Rail, and Todd Corporation of New Zealand.

In 1993, Bell South NZ, a subsidiary of BellSouth International, brought competition into the mobile communications market. The government licensed three bands of radio spectrum in a 1990 auction, and BellSouth and Telecom NZ made successful bids. The third spectrum was awarded to Telstra New Zealand. Interestingly, BellSouth International also holds 24.5 percent of an Australian telecommunications services operator, Optus, which is a competitor for Australia's telecommunications giant, Telstra.

Local and Long Distance Services

Once statutory barriers to entry into telecommunications network markets were removed, competition was introduced in both the long distance and local exchange services. Clear Communications, which started operations in 1991, was the first private competitor to Telecom NZ. By September 1995, Clear had gained a 22-percent market share in long distance and a 23-percent share of international service. Clear is now planning to provide local call services in central metropolitan areas.

Many New Zealanders feared that privatization and competition would erode service quality and lead to higher prices. Instead, they brought about increased investment in network facilities, improved productivity in the industry, lowered prices, and expanded markets. Telecom NZ has spent more than NZ$4.5 billion in its capital investment.

Table A10.2 Telecom services environment, New Zealand, 1997

Types of service	Price (US$)
Monthly charge, residential fixed line	51.73
Installation fee, residential fixed line	89.24
Local calls per minute	free
IDD[a] call to New York per minute	2.87
Cellular monthly charge, standard plan	48.59
Installation wait for residential line	1 day

a. IDD (international direct dial) is the cost of a phone call to New York using an incumbent carrier, Telecom New Zealand.

Source: Asian Wall Street Journal, 10 June 1997.

Long distance rates fell because of competition. Telecom NZ's off-peak tariff applied to the longest distance call declined by 37 percent from 47 cents per minute in 1987 to 29.4 cents per minute in 1995 (see table A10.1). Telecom NZ and Clear also offer a variety of discount packages.

Competition also improved efficiency. Telecom NZ was staffed by 26,500 persons when it separated from the New Zealand Post Office in 1987; the number of workers dropped to 8,600 by 1995. Although further reductions were planned, the number of workers increased to 9,100 in 1996.

Mobile Communications and Other Services

In 1995, mobile subscribers reached 3.9 million, or 10.8 sets per 100 people. Telecom NZ had been the only mobile service provider using the analog AMPS network until BellSouth successfully bid on the newly allocated radio spectrum for cellular service in 1993. BellSouth adopted the digital GSM system to compete with Telecom NZ. Telecom NZ now uses digital AMPS service as well.

Telecom NZ is trying to expand its business in Australia. It holds a 51-percent share in its Australian subsidiary, Pacific Star Communications, which offers enhanced facsimile transmission and other services. Cooperating with Bell Atlantic, Telecom NZ won the right to manage the provincial government's network in Queensland, Australia. Cooperating with Unisys of the United States, it also bid on contracts to automate the public records. In addition, Telecom NZ has equipment and management contracts in Fiji, Thailand, Western Samoa, and the Cook Islands.

1997 Environment

Table A10.2 summarizes New Zealand's environment for telecommunications services in 1997.

Philippines

Highlights from offer at WTO Negotiations on Basic Telecommunications (WTO 1997):

- Offers competition through commercial presence in the following services on a facilities basis for public use by means of all types of technologies except cable television and satellite: voice telephone, data transmission services, and cellular mobile telephone services in all market segments (local, long distance, and international). Will determine market access for the new entrants on the basis of public convenience and necessity.
- Limits foreign equity to 40 percent.
- Includes some regulatory principles.

Infrastructure and Policy

The telecommunications industry in the Philippines is unique. As early as 1981, there were 62 local telephone companies, 9 domestic telegraph and telex companies, and 3 data communications companies, all privately owned. Long distance service, however, has long been monopolized by the Philippine Long Distance Telephone Company (PLDT).

The telecommunications infrastructure in the Philippines is not well developed. Teledensity is low—4 main lines per 100 people—and the quality of service remains poor. The long-standing PLDT monopoly and the ineffective National Telecommunications Commission (NTC) are the main causes.

Deregulation in 1993 was intended to improve the telecommunications infrastructure and achieve nationwide teledensity of 8 by 1998 and 10 by 2000. The deregulation effort focused on two key provisions: (1) a mandated interconnection among the various telecommunications networks, and (2) the requirement for cellular phone companies to create 400,000 fixed lines and for international gateway operators to create 300,000 lines. Although only 1 of 11 companies, BayanTel, had finished installing the required lines as of March 1997 (see table A11.1), teledensity should reach 5 by the end of 1998. This growth prompted the government to raise the teledensity target to 14.4 by 2000.

The Philippines' offer in the WTO Telecommunications Pact indicates that some liberalization is under way.

Local and Long Distance Services

Following the deregulation of 1993, the Philippines' local and long distance markets were flooded with small local phone operators and at

Table A11.1 Main line installation requirement, Philippines, 1997

Carrier name	Total new lines required under deregulation plan	Lines installed as of 1 March 1997	Percentage of required lines installed
BayanTel	300,000	342,960	114.3
Digitel	300,000	169,786	56.6
PT&T/Capwire	300,000	90,500	30.2
Smart	700,000	186,386	26.6
Globe Telecom	700,000	118,506	16.9
Major/Philcom	300,000	45,502	15.2
Islacom	700,000	9,753	1.4
Piltel	400,000	na[b]	na
ETPI/TTPI	300,000	na[b]	na
PLDT[a]	0	761,375	na

na = not available.

a. PLDT is not required under the government's deregulation guidelines to install new phone lines.
b. Rollout scheduled for 1997.

Source: Asian Wall Street Journal, 10 June 1997.

least nine international phone operators. PLDT faced widespread competition, and its market share fell by almost 25 percent. Prestigious Filipino families established most of the new international phone systems, in conjunction with strong foreign partners. Since foreign ownership is limited to 40 percent, foreign telecommunications firms participate essentially as technical partners.

■ Globe Telecom aims to become the second long distance carrier; it is a subsidiary of Ayala, a property and manufacturing conglomerate, which holds 44 percent of its shares. Globe Telecom will be relying on the technical expertise of Singapore Telecom International, which holds a 38-percent stake. Globe aims to spend about US$1 billion in the next few years providing mobile, local, and international services in the Philippines. NTC approved Globe's entry as an international carrier in January 1994.

■ International Communications was set up by the Eugenio Lopez business clan in 1990 to operate satellite services. The Lopez group owns the country's largest broadcast network and controls Manila Electric. In 1993, Lopez purchased a telegram firm, Radio Communications Philippines, and a telephone company. Australia's Telstra International then became a partner in International Communications. In January 1994, NTC gave International Communications approval to become an international carrier.

■ Digital Telecommunications (Digitel) is 51 percent owned by shopping mall and manufacturing magnate John Gokongwei. Cable & Wireless

(United Kingdom) owns 28 percent of Digitel. Since 1990, the firm has been operating telephone systems in four central Luzon provinces and was awarded a P$40 billion (US$1.6 billion) contract to run and expand the government-owned system in northern Luzon. Digitel has expressed interest in developing a new national telephone network and offering international services. It is also seeking another foreign partner. NTC is expected to approve Digitel as an international carrier.

- Bell Telecommunications Philippines was set up by property magnate Francisco Ortigas and two other Filipino partners. BellSouth of the United States holds a 30-percent share of the firm.

The government's rule forcing PLDT to interconnect with other companies impelled two other companies to enter the long distance market: Eastern Telecommunications and Philippine Global Communications.

The next phase may be consolidation. Telecommunications Holding, for example, was set up in January 1994 to function as the holding company for the communications enterprises of Gokongwei and Lopez. Globe Telecom and International Communications have agreed to cooperate in establishing a national digital transmission network.

Faced with this onslaught of competition, PLDT has rapidly installed new telephone lines and is working to upgrade its services. Phone lines are finally reaching homes that had waited as long as three decades for basic telecommunications service. Many of the new lines are provided by rivals to PLDT.

Improvements in the quality of telephone service, however, have lagged. Thousands of customers, especially in congested Manila, now have a home or office phone line but still cannot make a call. Trunk lines are not adequate to handle the high volume of traffic.

The Philippines is also one of few countries to have experienced higher prices since telecommunications deregulation. In 1997, a basket of business telecom services cost US$4,000 per year, up 5 percent from 1996. By comparison, the same basket cost US$2,894 in Singapore and US$4,994 in the United States. Price increases can be attributed partly to the elimination of cross-subsidies. International deregulation lowered the price of international phone calls and, therefore, PLDT's revenue from international operations. PLDT compen-sated by raising its local charges. The basic subscription fee formerly permitted an unlimited number of local calls. In 1997, however, phone companies were allowed to increase their fixed-line services fee by 20 percent and charge local calls by the minute.

Mobile Communications Services

The NTC recently adopted a liberal stance in approving applications to operate both cellular mobile systems and international gateways. In 1992,

Table A11.2 Telecommunications services environment, Philippines, 1997

Types of service	Price (US$)
Monthly charge, residential fixed line	10.69
Installation fee, residential fixed line	7.96
Local calls	free
IDD[a] call to New York per minute	0.07
Cellular monthly charge, standard plan	31.29
Installation wait for residential line	14-365 days

a. IDD (international direct dial) is the cost of a phone call to New York using an incumbent carrier, PLDT.

Source: Asian Wall Street Journal, 10 June 1997.

Pilipino Telephone (Piltel), a PLDT subsidiary, dominated the cellular phone market. Express Telecommunications, which had a 20-percent market share, suddenly became an aggressive player when Philippine Communications Satellite purchased a 47-percent stake in it.

Since 1992, the NTC has approved three new applications for cellular operation: Globe Telecom, Isla Communications, and Smart Communications. Isla Communications, controlled by the Delgado family (a group with freight and fast-food enterprises), has US West as its technical partner. Smart Communications is owned by a group of Filipinos, who have a 63-percent stake in the company, and a local subsidiary of the First Pacific Group, an Indonesian company based in Hong Kong. Smart Communications began offering cellular services in February 1994. It is now negotiating to sell part of the company to NTT of Japan.

Newcomers are likewise rushing into the personal paging business, which is dominated by two firms: EasyCall and Pocketbell. In 1993, PLDT set up its own paging system, Beeper 150, and rapidly ate into market shares of the two incumbents. Since then, two additional firms have been given licenses to operate paging systems. Meanwhile, Express Telecommunications has signed a deal with EasyCall to offer customers a package that includes both a cellular phone and a pager.

Fraud is a serious problem for cellular phone operators. Many cellular phone operators do not have good systems for credit investigation and credit collection. Piltel and Globe together wrote off 85,000 fraudulent and 30,000 nonpaying customers, about 40 and 25 percent, respectively, of their total customers. Smart Communications, a relatively new operator, has developed a system to prevent this type of problem.

1997 Environment

Table A11.2 summarizes the Philippines' environment for telecommunications services in 1997.

Singapore

Highlights from offer at WTO Negotiations on Basic Telecommunications (WTO 1997):

- Commits to phase-in competition in facilities-based telecommunication services in April 2000, when up to two additional operators will be licensed; indicates that additional licenses will be granted thereafter.

- Offers open markets for mobile data, cellular telephony, and trunked radio services and for paging services as of April 2000.

- Commits to the provision of domestic and international resale of public-switched capacity (not including the connection of leased lines to public network) for most basic services, including voice, data, and ISDN.

- Limits foreign equity to 49 percent for facilities-based supply.

- Commits to the Reference Paper on regulatory principles.

Infrastructure and Policy

Singapore has established one of the most advanced telecommunications infrastructures in the world under a government-owned monopoly, the Telecommunications Authority of Singapore (TAS). Historically, TAS was responsible not only for regulation but also for operation in both the telecommunications and postal services. Main lines tripled in number from 1977 to 1989; by 1992, about 1.2 million main telephone lines—42 telephones per 100 inhabitants—had been installed. As early as 1987, a fiber-optic network linked all 26 of Singapore's telephone exchanges, and by 1994, the national telephone network was completely digitized. Virtually all homes and offices in Singapore are connected to a local network. The telecommunications industry is one of the most important industries in Singapore, accounting for more than 3 percent of Singapore's GDP in 1994.

In 1992, TAS was separated into three parts: Singapore Telecommunications PTE (Singapore Telecom or ST) is responsible for operating telecommunications services, Singapore Post is responsible for operating postal services, and TAS itself remains as the regulatory body. Singapore Telecom was granted monopoly rights until 1997 for cellular services and 2007 for fixed-line services, although these monopoly rights were modified by Singapore's offer at the WTO Telecommunications Pact to introduce competition in facilities-based telecom service in April 2000.

In October 1993, Singapore's government began privatizing ST with the goal of giving more flexibility to ST's management to invest and expand freely and aggressively overseas. Privatization of ST promotes the ultimate goal of establishing Singapore as a telecommunications hub of Asia.

Foreign participation in Singapore's telecommunications industry is quite limited. To limit foreign ownership of ST, the government issued three types of shares: Group A and B shares, about 50 percent of the shares offered, were reserved for Singaporeans; Group C shares have no limitation on investors' residency. Foreign ownership of facilities-based service operators is also subject to strict limitations. (Singapore promised to raise this limit to 41 percent in April 2000.) Foreign firms that are interested in operating in Singapore's telecommunications market are forced to cooperate with Singaporean firms.

Local and Long Distance Services

Singapore's telecommunications industry has long been monopolized by Singapore Telecom. Until early 1997, ST was the sole provider of almost all communications services—fixed line, international calls, cellular phone, radio paging, and mail services. Globalization of telecommunications, coupled with the WTO Telecommunications Pact, forced Singapore to change its telecommunications strategy from monopoly to competition. TAS announced that it will grant two additional licenses for operating basic telecommunications services; these licensees are expected to begin their operations in April 2000. Several firms, including foreign firms, reacted quickly to this announcement and started to prepare for the coming auction. To meet the 41-percent limit on foreign ownership of facilities-based service, foreign firms have begun establishing consortia with Singaporean firms. For example, NTT of Japan and British Telecom of the United Kingdom have formed a consortium with Singapore Technologies Telemedia (STT) and Singapore Power (SP). NTT and British Telecom each holds 20 percent of total shares, and the remaining 60 percent are divided evenly between STT and SP.

Singapore is already reaping the rewards of liberalization: international direct dial (IDD) rates have fallen by about 16 percent since January 1996. This decline can be attributed to threats from call-back operations and government pressure to keep IDD rates low. Overall, the prices charged for telecommunications services in Singapore are among the lowest in the world.[17]

---

17. The estimated annual cost of telecommunications services is US$2,894 for business users and US$1,720 for residential users (*Asian Wall Street Journal*, 10 June 1997).

Table A12.1 Phone subscription, Singapore, 1990-95

	Teledensity	Cellular subscription (thousands)
1990	39	51,000
1991	40	81,900
1992	42	120,000
1993	45	179,000
1994	45	235,630
1995	47	291,870

Source: International Telecommunication Union, 1995 World Telecommunication Development Report.

Mobile Communications Services

Mobile communications are becoming a necessity in Singapore. At the end of 1994, 27 percent of Singaporeans had pagers, and 7 percent had mobile phones (table A12.1 shows the growth in teledensity and cellular subscriptions between 1990 and 1995). Until 1997, ST was, through its subsidiary SingTel Mobile, the sole provider of all public cellular and radio paging services. In April 1997, MobileOne joined the cellular market and began offering wireless, cellular, and paging services. MobileOne is a consortium of four major business corporations: the Keppep Group, Singapore Press Holdings, Cable & Wireless, and Hong Kong Telecom. In the first three weeks of its operation, 35,000 cellular users, about 10 percent of the market share, signed up with MobileOne. MobileOne initially offered prices 10 percent lower than SingTel Mobile's. However, in late April 1997, SingTel Mobile lowered its monthly fee by up to 7 percent and its call charge by up to 20 percent. Thus, the average monthly bill for a medium-usage cellular subscriber has fallen about 40 percent in 1997 (*Asian Wall Street Journal,* 10 June 1997).

TAS has announced a plan to add two more operators in the cellular service market by April 2000. Foreign companies are among those interested in Singapore's cellular market. One consortium, led by a Singaporean telecommunications equipment manufacturer, includes Motorola, DDI (Japan's second-largest long distance operator), and AirTouch International (a US cellular operator). The government-funded firm, Singapore Technology Ventures, has announced its interest in entering the mobile and data communications market with technical assistance from BellSouth of the United States. British Telecom has shown interest in the cellular service market as well as the facilities-based service market.

To prepare for global competition, ST is adding new services. "GSM

Auto Roam," for example, enables a cellular customer to make calls from anywhere in the world where a GSM cellular telephone system is in operation.[18] At present, this service is available in Australia, the United Kingdom, Hong Kong, Denmark, and Switzerland, but not in the United States and Japan. Singapore Telecom's initiative could lead to further promotion of the GSM system throughout Asia.

Other Services

Singapore launched an ISDN service in 1989, and in 1990 became the first operator in the world to provide nationwide ISDN to all subscribers.[19] By mid-1993, ISDN served 18 major destinations, including Hong Kong, Japan, the United Kingdom, and the United States. In addition, Singapore's submarine cables and fiber-optic cables connect Singapore to all of its neighbors in the Association of Southeast Asian Nations (ASEAN).

Through its subsidiary, Singapore Telecom International, ST has been aggressively expanding overseas, searching out joint ventures with a variety of companies: cable services in Britain; data services in Thailand; cellular services in Vietnam, Sri Lanka, and the Philippines; and paging services in Indonesia and Hong Kong. In 1997, ST established a new data communications network for Myanmar. Previously, overseas telecommunications between Myanmar and the rest of the world had been restricted to telephone, telex, telegraph, and fax; however, with this new gateway, users in Myanmar now have access to e-mail, data transfer, and the internet. ST is now negotiating with Laos and Cambodia to set up similar data links.

Singapore's centrally directed effort to become a regional telecommunications center continues. The government plans to build a high-capacity network to deliver a range of on-line services, including cyber-shopping and high-speed internet access. Entertainment and educational sources, such as cable TV, interactive games, and government services, will be available through the network. Singapore authorities claim that by the end of 1997, 5,000 Singaporeans will be using the high-capacity network to go to the library, hang out in the video arcade, and send in payment for Singapore's famous fines (*Asian Wall Street Journal*, 10 June 1997).

An interesting question in the years ahead will be whether Singapore, with its centrally directed strategy, or Hong Kong, with its decentralized system, will emerge as the telecommunications hub of East Asia.

---

18. GSM is a digital cellular standard that is widely used in Asia and Europe.

19. ISDN is a technology for transmission of voice, data, and video over conventional copper telephone wire.

Table A12.2 Telecommunications services environment, Singapore, 1997

Types of service	Price (US$)
Monthly line fee, residential fixed line	5.83
Installation fee, residential fixed line	20.98
IDD[a] call to Ner York per minute	0.84
Celluar monthly charge, standard plan	27.97
Installation wait for residential line	5 days

a. IDD (International Direct Dial) is the cost of a phone call using an incumbent carrier, Singapore Telecom.

Source: Asian Wall Street Journal, 10 June 1997.

1997 Environment

Table A12.2 summarizes Singapore's environment for telecommunications services in 1997.

United States

Highlights from offer at WTO Negotiations on Basic Telecommunications (WTO 1997):

■ Commits to open markets for essentially all basic telecommunications services (facilities-based and resale) for all market segments (local, long distance, and international), including unrestricted access to common carrier radio licenses for operators that are indirectly foreign-owned. Offer also covers satellite-based services, cellular telephony, and other mobile services.

■ Limitations on market access include no issuance of radio licenses to operators with more than 20 percent direct foreign ownership, and Comsat retains exclusive rights to links with Intelsat and Inmarsat satellite capacity.

■ Commits to the Reference Paper on regulatory principles.

Infrastructure and Policy

In the United States, telecommunications infrastructure has achieved one of the highest levels in the world: by 1995, 165 million main lines had been installed, on 63 lines per 100 people. US telecommunications operators provide high-quality domestic communications services and extend their operations worldwide.

Historically, telecommunications services were provided by a single private company, AT&T, which dominated the market. The US Department of Justice tried several times to break up the AT&T monopoly; finally, the AT&T Consent Decree (known as Modified Final Judgment or MFJ) resulted in the breakup of AT&T in 1984. AT&T's local services operators became independent Regional Bell Operating Companies (RBOCs), but it kept its long distance and international service as well as its manufacturing operations.

In the United States, there are two main regulatory systems. The first of these is a single body, the Federal Communications Commission (FCC). The FCC was established by the Communications Act of 1934 with the goal of encouraging competition in all communications markets and protecting the public interest. It regulates interstate and international communications by radio, television, wire, satellite, and cable. The second system is a set of regulatory bodies, state utility boards, which regulate local communications.

The Telecommunications Act of 1996

The Telecommunications Act of 1996 (1996 Act) was the first major change in telecommunications law since 1934. Its objective was to establish a competitive environment in the telecommunications industry by (a) opening local exchange markets to competitive entry; (b) promoting more competition in the long distance market; and (c) reforming the US universal service system. This section mainly discusses the issues of universal service; competition in local and long distance markets is discussed in the next section.

The primary argument against introducing competition in local phone markets is that competitive local markets will not foster universal service.[20] Historically, universal basic phone service has been supported by cross-subsidies from long distance (and international) service, and by cross-subsidies among different clusters of local service. The FCC recognizes that the current cross-subsidy system acts as a barrier to the introduction of competition in local markets. Understandably, incumbent local operators do not want to be left with a universal service obligation to their unprofitable customers while new entrants capture their lucrative customers. To avert this possibility, incumbent RBOCs want to charge a fairly high interconnection or access charge to use their networks. However, high access charges make it difficult for new entrants to start an

---

20. Basically, the term *universal service* means reasonably priced access by all interested households to all modes of telecommunication.

operation by leasing local network capacity (the so-called resale business). The FCC has proposed interconnection rules that lean toward encouraging new entrants by setting low access charges. The RBOCs have vigorously resisted such rules.

Expanding the concept of universal service, the 1996 Act requires the FCC to guarantee not only affordable telephone services to all Americans but also internet access to all classrooms, libraries, and health care clinics. These goals are difficult to accomplish without some type of federal funding system. In May 1997, the FCC proposed a new universal annual fund to replace the current cross-subsidy regime. Under this universal fund system, local phone operators, known in the new jargon as local exchange companies (LECs), can charge a subscriber a line charge, in addition to a monthly subscription fee, of up to US$3.50 a month for a first line and US$5.00 for a second line. Also, LECs can raise their monthly subscription fees. In the aggregate, it is expected that higher local charges will be offset by cheaper long distance calls.

Those opposed to the universal fund system find two big problems in the FCC proposal. They question whether reducing access charges will in fact lower the cost of long distance calls. They also point out that customers who make few long distance calls will only see an increase in their local service charges with no corresponding benefit from cheaper long distance calls. Since long distance calls are income-elastic, this system will hurt the poor and benefit the rich. The FCC responded that US$3.50 is the maximum subscription fee that LECs can charge for the first line, and that monthly local charges are not likely to rise very much. The FCC also reported that it would examine the LECs' pricing system to determine whether a price increase is necessary to maintain universal service. Nevertheless, the FCC's lynchpin argument is that the universal fund makes possible the end of the cross-subsidy system.

Local and Long Distance Services

In 1984, under the Modified Final Judgment, AT&T was separated from seven local telecommunications companies, the RBOCs or "Baby Bells": Nynex of New York, Bell Atlantic of Philadelphia, BellSouth of Atlanta, Southwestern Bell of Texas, US West of Denver, Ameritech of Chicago, and Pacific Telesis of San Francisco.[21] AT&T retained control over its manufacturing operations and was forced to compete in the long distance and international service markets. Sharp competition in long distance and international service emerged. Table A13.1 summarizes the

---

21. LECs consist of RBOCs, GTE, and other telecommunications firms that possess local telephone exchanges.

Table A13.1 Changes in long distance market shares, 1984-95

	1984	1985	1986	1987	1988	1989	1990	1991	1992	1993	1994	1995
Long distance calls												
Industry total (billion minutes)	na	167.1	183.1	215.7	244.6	277.1	307.4	328.0	349.7	371.2	401.4	432.0
AT&T (billion minutes)	na	133.3	140.6	155.3	167.6	179.9	192.6	204.0	211.7	223.6	235.1	244.1
Share of AT&T (percentages)	na	79.8	76.8	72.0	68.5	64.9	62.7	62.2	60.5	60.2	58.6	56.5
Subscribed telephone lines												
Industry total (million)	na	na	na	121.5	124.4	128.5	132.4	135.3	138.7	142.8	148.5	152.6
AT&T (million)	na	na	na	101.7	100.2	99.4	100.1	101.5	101.2	101.7	104.0	101.1
MCI (million)	na	na	na	10.0	12.1	15.1	17.4	18.3	20.2	21.8	22.0	23.9
Sprint (million)	na	na	na	5.8	7.2	8.2	8.7	8.4	8.9	9.2	9.5	9.8
Share of AT&T (percentages)	na	na	na	83.7	80.6	77.4	75.6	75.0	73.0	71.2	70.0	66.3
Share of MCI (percentages)	na	na	na	8.2	9.8	11.7	13.2	13.5	14.5	15.3	14.8	15.7
Share of Sprint (percentages)	na	na	na	4.8	5.8	6.4	6.6	6.2	6.4	6.5	6.4	6.4
Toll service revenues												
Industry total (million US$)	51,156	54,815	57,468	58,519	62,600	66,024	66,792	68,558	71,983	75,290	80,726	83,782
AT&T (million US$)	34,935	36,770	36,514	35,219	35,407	34,549	33,880	34,384	35,495	35,731	37,166	38,069
MCI (million US$)	1,866	2,532	3,663	4,334	5,410	6,884	7,392	8,266	9,719	10,947	11,715	12,924
Sprint (million US$)	1,052	1,509	2,132	2,592	3,405	4,320	5,041	5,378	5,658	6,139	6,805	7,277
Share of AT&T (percentages)	68.3	67.1	63.5	60.2	56.6	52.3	50.7	50.2	49.3	47.5	46.0	45.4
Share of MCI (percentages)	3.6	4.6	6.4	7.4	8.6	10.4	11.1	12.1	13.5	14.5	14.5	15.4
Share of Sprint (percentages)	2.1	2.8	3.7	4.4	5.4	6.5	7.5	7.8	7.9	8.2	8.4	8.7

na = not available.

Source: Federal Communications Commission, Long Distance Market Shares (1996, 3Q).

Table A13.2 Long distance prices, United States, 1985-95

	1985	1986	1987	1988	1989	1990	1991	1992	1993	1994	1995
Long distance calls											
Industry total (cents/minute)	32.8	31.4	27.1	25.6	23.8	21.7	20.9	20.6	20.3	20.1	19.4
AT&T (cents/minute)		27.6	26.0	22.7	21.1	19.2	17.6	16.9	16.8	16.0	15.8	15.6

Note: As a proxy for unit prices, toll service revenue is divided by long distance call minutes.

Source: Federal Communications Commission, Long Distance Market Shares (1996, 3Q).

changes in the long distance market over the past decade. In 1995, AT&T held 45 percent of the long distance market (in revenue terms), while MCI held 15 percent and Sprint 9 percent; the remaining 30 percent was held by hundreds of smaller operators, including LDDS communications, a potential fourth major competitor.

Competition has lowered the price of long distance calls. For example, the average price of a 10-minute call from Chicago to Atlanta dropped from US$5.65 to US$2.37 between 1985 and 1992. (Table A13.2 shows price changes between 1985 and 1995.)

While the long distance and international service markets have undergone substantial liberalization, local markets remain virtually monopolized. Only six states permit competition in the local phone market. For example, in New York, Time Warner Communications is approved to provide telephone service.

Major long distance operators, which are paying more than US$20 billion in annual access charges to LECs, want to lower these high fees. They are forging cheaper alliances with various cable TV operators. For example, Sprint and its three cable TV partners, including Telecommunications (TCI, the biggest cable TV operator in the United States), will form a new telecommunications company to provide multimedia services.

A new high-powered network, the fiber-optic-based metropolitan area network (MAN), is also emerging. MANs are aimed at business customers who need high-powered cable for inter- or intra-office networking. The revenue from these systems has been growing at about 22 percent a year (Noam 1995). AT&T has pioneered the MAN concept. But AT&T spends only US$1 billion annually for local network construction, while the Baby Bells and GTE together spend about US$24 billion for local network construction and maintenance. These figures indicate both the expense of network construction and the disparity in infrastsructure outlays by RBOCs and long distance operators.

Hence, for many potential entrants in local phone service, the resale route seems an inviting way to avoid the huge initial cost of building a new local network. MCI and Sprint used the resale strategy successfully in the 1970s and 1980s to enter the long distance market. To foster the resale business, the 1996 Act requires LECs to allow their competitors to interconnect wih their local network on a nondiscriminatory basis. The *quid pro quo* is that each LEC will be allowed to offer long distance service once its own local market becomes effectively competitive.

To foster interconnection, the FCC's new pricing rule was designed to create a price-equals-incremental-cost system for interconnection access to the local phone market. This Total Service Long Run Incremental Cost (TSLRIC) concept also allows for a reasonable share of forward-looking joint and common costs. TSLRIC is designed to calculate the cost based on building a network and operating it with the most advanced technology.

Potential new entrants like the TSLRIC pricing rule. Echoing the FCC, they contend that the TSLRIC system rule leads to prices that would prevail in a competitive market. Not surprisingly, most LECs oppose the TSLRIC system. They argue that the TSLRIC system denies LECs the opportunity to recover the high costs of their existing infrastructure. LECs also argue that the TSLRIC system is unfair, because not all existing networks can use the most advanced technology to expand capacity. Hence, in their view, the TSLRIC system will generate operating losses for many LECs and discourage new infrastructure investment. The LECs' arguments are based on five key points: (a) the TSLRIC system will not give an adequate incentive for new entrants (and many LECs) to build their own facilities; (b) it is highly subjective and will be costly to implement and audit; (c) it does not reflect regional differences; (d) it precludes an adequate contribution to joint and common costs; and, finally, (e) it does not allow the LECs to recover a return on their historic investment costs.

In September 1996, several LECs and state utility boards sued the FCC over its pricing rule. The Eighth Circuit Court issued a preliminary injunction in favor of the petitioners, remarking that the FCC's proposed pricing rule ignores the power and ability of state commissions to check LECs' discriminatory or anticompetitive pricing rules. The court concluded that competition in the local phone market can be enhanced without the new FCC pricing rule.

AT&T and 34 other companies filed a countersuit in the Supreme Court to overturn the Eighth Circuit Court's decision. They contended that the FCC's new rules carry out the intent of the 1996 Act. Without this new rule, they argue, the LECs will use their power to prevent effective competition in local markets. Alternative pricing rules proposed by the LECs will delay the introduction of competition in local markets. In fact, they say, local markets may go back to de facto monopoly if the FCC's new rule is not implemented. In November 1996, the Supreme Court lifted the Eighth Circuit's preliminary injunction. When the case was argued on its merits in July 1997, the Circuit Court overruled the

FCC's pricing rule. The court said that establishing a pricing rule for interconnection exceeded the FCC's jurisdiction under the 1996 Act.

Leaving aside the legal battle, expected profits from the resale business are not particularly inviting. To enter a local phone market, it often makes more sense to acquire a local phone operator that owns a local network. Mergers and acquisitions among telecommunications firms are not new. In 1994, Sprint, the third-largest long distance company, announced that two major European firms, Deutsche Telekom and France Telecom, had each purchased a 20-percent stake in it. In 1996, MCI and British Telecom announced a merger plan by which British Telecom would acquire MCI for US$21 billion. While this deal between MCI and British Telecom was waiting for FCC approval, MCI received two new offers. Worldcom, an internet provider, offered a $37 billion stock offer; GTE, a local, long distance, and internet provider, offered a $28 billion cash offer. As of December 1997, Worldcom is likely to merge with MCI, although FCC approval has not yet been gained.

Domestically, in April 1997 the Justice Department approved a merger plan by which Bell Atlantic will acquire Nynex for US$23 billion and create a giant local phone operator. Because the new company will account for about 25 percent of the US local phone market, the FCC required Bell Atlantic to agree to meet four conditions of approval:

- that it negotiate with competitors over standards involving the delivery of services or the connection of equipment;

- that it develop a uniform way for competitors to order services electronically for their new customers;

- that it offer installment payments for certain charges that smaller-scale competitors face when signing up new customers or connecting their equipment to the local phone network; and

- that it not count the historic cost of its network when deciding how much to charge competitors for interconnections (*Washington Post*, 20 July 1997).

In this agreement, the pricing rule seems highly similar to the TSLRIC concept. Nevertheless, some telecommunications competitors are still skeptical about the consequences of the merger.

Other merger plans have been announced and not carried out, notably between AT&T and SBC. SBC is Texas Bell and had already acquired Pacific Telesis in California for US$17 billion. The AT&T and SBC merger would have been twice the size of the previous largest merger in corporate history and would have created the biggest concentration of assets in the telecommunications industry. Both AT&T and SBC recognized the difficulties of making this merger acceptable to regulators. Attempts to put a procompetitive face on the deal were both painful to SBC and inadequate for critics (including Reed Hundt, chairman of the FCC) (*Wall Street Journal,* 30 June 1997).

Although the Bell Atlantic/Nynex mega-merger was the only one planned for 1997, the trend in the US telecommunications market seems to be toward consolidation rather than head-to-head competition. Critics contend that the 1996 Act sets telecommunications firms free to reorganize as giant oligopolies—with less, not more, competition in the US markets.

Mobile Communications Services

Ten years ago, the cellular telephone business barely existed; by 1993, over 30 million cellular phones were in use in some 70 countries—13 million of them in the United States. The US cellular market has been competitive regionally since its start in the early 1980s. This situation reflects the FCC policy of giving two operating licenses for each region, one to each of the established local telephone companies and one to a new independent rival.

AT&T became an aggressive new entrant in the cellular markets after it paid US$11.5 billion to acquire one-third of McCaw Cellular Communications, the largest cellular operator in the United States. McCaw will be renamed AT&T Wireless Services. The union of AT&T and McCaw created a threat to other rivals and spawned a wave of new alliances. Bell Atlantic and Nynex announced that their combined cellular subsidiaries will form a nationwide cellular firm. US West and AirTouch Communications (a California-based cellular phone company and the former cellular arm of Pacific Telesis) plan to join the Bell Atlantic/Nynex deal. Sprint, meanwhile, is attempting to create a nationwide wireless network under an alliance already formed with three cable TV operators. All these deals remain to be approved by the FCC. However, they underline the erosion of the walls that once separated local, long distance, and wireless services.

Several foreign telecommunications firms, especially British operators, are interested in the profitable US cellular market, but their scope of operation is severely limited by the 25-percent ceiling on foreign ownership of US firms that hold a radio-based license. This restriction, originally imposed for security reasons during World War I, had become a protectionist device by the 1980s. Under the 1996 Act and the US offer at the WTO Telecommunications Pact, it will be removed.

The US telecommunications industry is rushing into a new generation of cellular technology, known as PCS.[22] PCS will challenge the

---

22. For a description of the PCS system, see the Overview section.

monopolies of the seven regional Bell operating companies. By 1993, the FCC had granted experimental PCS licenses to over 200 companies. The FCC has set aside 160 MHz of radio spectrum for the national PCS system, more than three times the 50 MHz allocated to existing cellular companies. In July 1994, it conducted its first auction of PCS licenses intended for paging and advanced messaging services.[23] Six companies purchased nationwide narrowband PCS licenses for a surprising US$617 million. In addition, hundreds of smaller bidders put up US$217 million for the 594 regional and local licenses.

The FCC is auctioning 99 broadband PCS licenses, which will be used to provide wireless voice and data transmission in direct competition with existing cellular systems.[24] It plans to grant PCS licenses for up to seven competitors in each PCS region. The rush of alliance formations discussed above was in part preparation for the coming auction.

Nextel is trying to create yet another type of nationwide wireless communications network using the radio dispatch system called Specialized Mobile Radio (SMR). This network would compete both with cellular telephone systems and PCS. Nextel's SMR is already ahead of the PCS network. To build its system, Nextel acquired one of its competitors, Dial Page, a radio communications network operator for truck and taxicab dispatching in South Carolina. Nextel also invested in OneComm, a Denver-based paging company. Nextel plans to purchase mobile nationwide communications licenses from Motorola. These actions extend Nextel's airwave rights to cover 85 percent of the US population. In February 1994, MCI announced the possible acquisition of 17 percent of Nextel, but the deal was suspended when Motorola entered the picture. Motorola agreed to sell its entire portfolio of radio-dispatch properties for US$1.7 billion in Nextel stock, which gives Motorola a larger stake (21 percent) in Nextel than MCI would have held.

By 2000, analysts predict, Nextel will have 15 percent of total mobile communications subscribers, while PCS systems will account for about 35 percent of total subscribers. The remaining 50 percent of subscribers will use current mobile technology.

Movements toward an Interactive Multimedia Network

In October 1993, Bell Atlantic announced a merger with Telecommunications, the nation's largest cable company. The deal, valued at approxi-

---

23. Paging Network, the largest paging company, bid US$197 million for three nationwide licenses; McCaw Cellular, the fifth-largest in the paging field, gained two for US$160 million; and Mobile Telecommunications won two licenses for US$127 million. BellSouth Wireless, AirTouch Paging, and Pagemart II each won one license.

24. See the Overview section.

mately US$33 billion, stunned industry observers. Although this deal was later withdrawn, it characterizes the search for ways to create a multimedia network—a component of the much-discussed information superhighway. Before the announcement, 1993 had already seen 10 mergers between US communications and media companies.

In the midst of the industry's reorganization, in early 1994, the Clinton administration hammered out a new policy toward multimedia networks. Telephone and cable companies have traditionally been separated by law and barred from entering each other's markets. The key concept in the new plan is "competitive reciprocity." A cable TV company can begin to offer local phone services as long as it faces direct competition in its own market from another video provider, such as the local phone company. Conversely, a local phone company can begin to sell TV service over its lines, once other telephone competitors have entered its market.

In mid-1994, Bell Atlantic became the first telecommunications company to win FCC approval to provide an interactive television service in competition with local cable television companies. Specifically, the company was authorized to provide "video dial tone service" to 38,000 households in a New Jersey township. Other telecommunications companies have similar applications, and approval seems likely for all. However, telephone companies are still forbidden to own the content of the programs they distribute. Bell Atlantic therefore teamed up with an independent programmer.[25] Under the 1996 Act, telephone companies and cable TV companies can potentially compete in all lines of business.

Other Services

In 1993, the telecommunications and the computer industries rushed into the market of "personal communicators," handheld devices that combine some of the functions of a personal computer with those of a mobile telephone, pager, and/or fax machine. Consumer response has so far been slow.

Apple Computer's "Newton," which was launched in 1993, fell far short of its sales targets of several million and sold fewer than 100,000 units. Eo, a Silicon Valley venture that brought the first personal communicator to the market, closed down in 1994 after AT&T, its largest investor, refused to provide additional funding. Intel ended its effort to develop chips for personal communicators. Other large companies, such

---

25. Likewise, several RBOCs—Ameritech, BellSouth, and Southwestern Bell—announced in August 1994 the formation of a joint venture with Walt Disney, to offer cable and interactive programming services.

as IBM, Compaq, and Motorola, have also delayed their programs. Nonetheless, many people remain convinced that the personal communicators will eventually become a mass market item (*The Financial Times*, 9 August 1994).

References

Blanchard, Carl. 1994. Telecommunications Regulations in New Zealand. *Telecommunications Policy* (March).
Chen, Bo-Shoe. 1994. Taiwan: Reform at a Snail's Pace. *Telecommunications Policy* (April).
Drake, William J., ed. 1995. *The New Information Infrastructure: Strategies for U.S. Policy.* New York: Twentieth Century Fund.
East-West Center. 1992. *Telecommunications: Southeast Asia's New Revolution.* Asia Pacific Briefing Paper.
Globerman, Steven, Tae Hoon Oum, and W. T. Stanbury. 1993. Competition in Public Long-Distance Telephone Markets in Canada. *Telecommunications Policy* (May/June).
Jussawalla, Meheroo. 1993. Telecommunications and Regional Interdependence in South East Asia. The Fletcher Forum of World Affairs.
Law, Carl Edgar. 1994. Asia-Pacific Telecommunications to 2000. Financial Times Management Reports. London: Financial Times.
Lee, Paul S. N. 1993. Hong Kong as a Communications Hub. *Telecommunications Policy* (September/October).
Lee, Paul S. N. 1994. China's Role in Hong Kong's Telecommunications Deregulation. *Telecommunications Policy* (April).
Noam, Eli M. 1995. Beyond Telecommunications Liberalization: Past Performance, Present Hype, and Future Direction. In *The New Information Infrastructure: Strategies for U.S. Policy,* ed. by William J. Drake. New York: Twentieth Century Fund.
Tan, Zixiang. 1994. Challenges to the MPT's Monopoly. *Telecommunications Policy* (April).
Tyson, Laura D'Andrea. 1992. *Who's Bashing Whom? Trade Conflict in High-Technology Industries.* Washington: Institute for International Economics.
Ure, John. 1994. Telecommunications, with Chinese Characteristics. *Telecommunications Policy* (April).
US Trade Representatives (USTR). 1993. *National Trade Estimate Report on Foreign Trade Barriers.* Washington: Government Printing Office.
Warwick, William. A Review of AT&T's Business History in China. *Telecommunications Policy* (April).
Williamson, Maurice. 1994. Recent Developments in New Zealand CommunicationsIndustries. Presentation delivered at the Embassy of New Zealand, Washington (1 December).
World Bank. 1994. *Telecommunications Sector Reform in Asia.* World Bank Discussion Papers 232. Washington: World Bank.
World Trade Organization (WTO). 1997. The WTO Negotiations on Basic Telecommunications: Informal Summary of Commitments and M.F.N. Exemptions. http://www.wto.org/new/bt-summ3.htm (6 March).

Other Publications from the
Institute for International Economics

POLICY ANALYSES IN INTERNATIONAL ECONOMICS Series

Trade Policy in the 1980s
William R. Cline, editor/*1983*
(out of print)   ISBN paper 0-88132-031-5   810 pp.

Subsidies in International Trade
Gary Clyde Hufbauer and Joanna Shelton Erb/*1984*
                             ISBN cloth 0-88132-004-8   299 pp.

International Debt: Systemic Risk and Policy Response
William R. Cline/*1984*   ISBN cloth 0-88132-015-3   336 pp.

Trade Protection in the United States: 31 Case Studies
Gary Clyde Hufbauer, Diane E. Berliner, and Kimberly Ann Elliott/*1986*
(out of print)   ISBN paper 0-88132-040-4   371 pp.

Toward Renewed Economic Growth in Latin America
Bela Balassa, Gerardo M. Bueno, Pedro-Pablo Kuczynski,
and Mario Henrique Simonsen/*1986*
(out of stock)   ISBN paper 0-88132-045-5   205 pp.

Capital Flight and Third World Debt
Donald R. Lessard and John Williamson, editors/*1987*
(out of print)   ISBN paper 0-88132-053-6   270 pp.

The Canada-United States Free Trade Agreement: The Global Impact
Jeffrey J. Schott and Murray G. Smith, editors/*1988*
                             ISBN paper 0-88132-073-0   211 pp.

World Agricultural Trade: Building a Consensus
William M. Miner and Dale E. Hathaway, editors/*1988*
                             ISBN paper 0-88132-071-3   226 pp.

Japan in the World Economy
Bela Balassa and Marcus Noland/*1988*
                             ISBN paper 0-88132-041-2   306 pp.

America in the World Economy: A Strategy for the 1990s
C. Fred Bergsten/*1988*   ISBN cloth 0-88132-089-7   235 pp.
                             ISBN paper 0-88132-082-X   235 pp.

Managing the Dollar: From the Plaza to the Louvre
Yoichi Funabashi/*1988, 2d ed. 1989*
                             ISBN paper 0-88132-097-8   307 pp.

United States External Adjustment and the World Economy
William R. Cline/*May 1989*   ISBN paper 0-88132-048-X   392 pp.

Free Trade Areas and U.S. Trade Policy
Jeffrey J. Schott, editor/*May 1989*
                             ISBN paper 0-88132-094-3   400 pp.

Dollar Politics: Exchange Rate Policymaking in the United States
I. M. Destler and C. Randall Henning/*September 1989*
(out of print)   ISBN paper 0-88132-079-X   192 pp.

Latin American Adjustment: How Much Has Happened?
John Williamson, editor/*April 1990*
                             ISBN paper 0-88132-125-7   480 pp.

The Future of World Trade in Textiles and Apparel
William R. Cline/*1987, 2d ed. June 1990*
                             ISBN paper 0-88132-110-9   344 pp.

**Completing the Uruguay Round: A Results-Oriented Approach
to the GATT Trade Negotiations**
Jeffrey J. Schott, editor/*September 1990*
                             ISBN paper 0-88132-130-3   256 pp.

Economic Sanctions Reconsidered (in two volumes)
Economic Sanctions Reconsidered: Supplemental Case Histories
Gary Clyde Hufbauer, Jeffrey J. Schott, and Kimberly Ann Elliott/*1985, 2d ed. December 1990*
                             ISBN cloth 0-88132-115-X   928 pp.
                             ISBN paper 0-88132-105-2   928 pp.

Economic Sanctions Reconsidered: History and Current Policy
Gary Clyde Hufbauer, Jeffrey J. Schott, and Kimberly Ann Elliott/*December 1990*
ISBN cloth 0-88132-136-2 288 pp.
ISBN paper 0-88132-140-0 288 pp.

Pacific Basin Developing Countries: Prospects for the Future
Marcus Noland/*January 1991* ISBN cloth 0-88132-141-9 250 pp.
(out of print) ISBN paper 0-88132-081-1 250 pp.

Currency Convertibility in Eastern Europe
John Williamson, editor/*October 1991*
ISBN paper 0-88132-128-1 396 pp.

International Adjustment and Financing: The Lessons of 1985-1991
C. Fred Bergsten, editor/*January 1992*
ISBN paper 0-88132-112-5 336 pp.

North American Free Trade: Issues and Recommendations
Gary Clyde Hufbauer and Jeffrey J. Schott/*April 1992*
ISBN paper 0-88132-120-6 392 pp.

Narrowing the U.S. Current Account Deficit
Allen J. Lenz/*June 1992*
(out of print) ISBN paper 0-88132-103-6 640 pp.

The Economics of Global Warming
William R. Cline/*June 1992* ISBN paper 0-88132-132-X 416 pp.

U.S. Taxation of International Income: Blueprint for Reform
Gary Clyde Hufbauer, assisted by Joanna M. van Rooij/*October 1992*
ISBN cloth 0-88132-178-8 304 pp.
ISBN paper 0-88132-134-6 304 pp.

Who's Bashing Whom? Trade Conflict in High-Technology Industries
Laura D'Andrea Tyson/*November 1992*
ISBN paper 0-88132-106-0 352 pp.

Korea in the World Economy
Il SaKong/*January 1993* ISBN paper 0-88132-106-0 328 pp.

Pacific Dynamism and the International Economic System
C. Fred Bergsten and Marcus Noland, editors/*May 1993*
ISBN paper 0-88132-196-6 424 pp.

Economic Consequences of Soviet Disintegration
John Williamson, editor/*May 1993*
ISBN paper 0-88132-190-7 664 pp.

Reconcilable Differences? United States-Japan Economic Conflict
C. Fred Bergsten and Marcus Noland/*June 1993*
ISBN paper 0-88132-129-X 296 pp.

Does Foreign Exchange Intervention Work?
Kathryn M. Dominguez and Jeffrey A. Frankel/*September 1993*
ISBN paper 0-88132-104-4 192 pp.

Sizing Up U.S. Export Disincentives
J. David Richardson/*September 1993*
ISBN paper 0-88132-107-9 192 pp.

NAFTA: An Assessment
Gary Clyde Hufbauer and Jeffrey J. Schott/*rev. ed. October 1993*
ISBN paper 0-88132-199-0 216 pp.

Adjusting to Volatile Energy Prices
Philip K. Verleger, Jr./*November 1993*
ISBN paper 0-88132-069-2 288 pp.

The Political Economy of Policy Reform
John Williamson, editor/*January 1994*
ISBN paper 0-88132-195-8 624 pp.

Measuring the Costs of Protection in the United States
Gary Clyde Hufbauer and Kimberly Ann Elliott/*January 1994*
ISBN paper 0-88132-108-7 144 pp.

The Dynamics of Korean Economic Development
Cho Soon/*March 1994*
ISBN paper 0-88132-162-1 272 pp.

The Trading System After the Uruguay Round
John Whalley and Colleen Hamilton/*July 1996*
ISBN paper 0-88132-131-1

224 pp.

Private Capital Flows to Emerging Markets After the Mexican Crisis
Guillermo A. Calvo, Morris Goldstein, and Eduard Hochreiter/*September 1996*
ISBN paper 0-88132-232-6

352 pp.

The Crawling Band as an Exchange Rate Regime:
Lessons from Chile, Colombia, and Israel
John Williamson/*September 1996*
ISBN paper 0-88132-231-8

192 pp.

Flying High: Civil Aviation in the Asia Pacific
Gary Clyde Hufbauer and Christopher Findlay/*November 1996*
ISBN paper 0-88132-231-8

232 pp.

Measuring the Costs of Visible Protection in Korea
Namdoo Kim/*November 1996*
ISBN paper 0-88132-236-9

112 pp.

The World Trading System: Challenges Ahead
Jeffrey J. Schott/*December 1996*
ISBN paper 0-88132-235-0

350 pp.

Has Globalization Gone Too Far?
Dani Rodrik/*March 1997*   ISBN cloth 0-88132-243-1

128 pp.

Korea-United States Economic Relationship
C. Fred Bergsten and Il SaKong, editors/*March 1997*
ISBN paper 0-88132-240-7

152 pp.

Summitry in the Americas: A Progress Report
Richard E. Feinberg/*April 1997*
ISBN paper 0-88132-242-3

272 pp.

Corruption and the Global Economy
Kimberly Ann Elliott/*June 1997*
ISBN paper 0-88132-233-4

256 pp.

Regional Trading Blocs in the World Economic System
Jeffrey A. Frankel/*October 1997*
ISBN paper 0-88132-202-4

346 pp.

Sustaining the Asia Pacific Miracle: Environmental Protection and
Economic Integration
André Dua and Daniel C. Esty/*October 1997*
ISBN paper 0-88132-250-4

232 pp.

Trade and Income Distribution
William R. Cline/*November1997*
ISBN paper 0-88132-216-4

296 pp.

Global Competition Policy
Edward M. Graham and J. David Richardson/*December 1997*
ISBN paper 0-88132-166-4

616 pp.

Unfinished Business: Telecommunications after the Uruguay Round
Gary Clyde Hufbauer and Erika Wada/*December 1997*
ISBN paper 0-88132-257-1

272 pp.

SPECIAL REPORTS

1 **Promoting World Recovery: A Statement on Global Economic Strategy**
   by Twenty-six Economists from Fourteen Countries/*December 1982*
   (out of print)   ISBN paper 0-88132-013-7

45 pp.

2 **Prospects for Adjustment in Argentina, Brazil, and Mexico:**
   Responding to the Debt Crisis   (out of print)
   John Williamson, editor/*June 1983*
   ISBN paper 0-88132-016-1

71 pp.

3 **Inflation and Indexation: Argentina, Brazil, and Israel**
   John Williamson, editor/*March 1985*
   ISBN paper 0-88132-037-4

191 pp.

4 **Global Economic Imbalances**
   C. Fred Bergsten, editor/*March 1986*
   ISBN cloth 0-88132-038-2
   ISBN paper 0-88132-042-0

126 pp.
126 pp.

WORKS IN PROGRESS

The US - Japan Economic Relationship
C. Fred Bergsten, Marcus Noland, and Takatoshi Ito
China's Entry to the World Economy
Richard N. Cooper
Liberalizing Financial Services
Wendy Dobson and Pierre Jacquet
Economic Sanctions After the Cold War
Kimberly Ann Elliott, Gary C. Hufbauer and Jeffrey J. Schott
Trade and Labor Standards
Kimberly Ann Elliott and Richard Freeman
The Asian Financial Crisis
Morris Goldstein
Forecasting Financial Crises: Early Warning Signs for Emerging Markets
Morris Goldstein and Carmen Reinhart
Prospects for Western Hemisphere Free Trade
Gary Clyde Hufbauer and Jeffrey J. Schott
Agricultural Trade Policy
Tim Josling
Trade Practices Laid Bare
Donald Keesing
The Future of U.S. Foreign Aid
Carol Lancaster
The Economics of Korean Unification
Marcus Noland
International Trade and Investment
Catherine L. Mann
Foreign Direct Investment in Developing Countries
Theodore Moran
Globalization, the NAIRU, and Monetary Policy
Adam Posen
The Case for Trade: A Modern Reconsideration
J. David Richardson
Measuring the Cost of Protection in China
Zhang Shuguang, Zhang Yansheng, and Wan Zhongxin
Real Exchange Rates for the Year 2000
Simon Wren-Lewis and Rebecca Driver

Canadian customers   RENOUF BOOKSTORE
can order from   5369 Canotek Road, Unit 1, Ottawa, Ontario K1J 9J3, Canada
the Institute or from: Telephone: (613) 745-2665   Fax: (613) 745-7660

Visit our website at: http://www.iie.com   **E-mail address: orders@iie.com**